Praise for *The Seed Farmer*

Seed saving is the next frontier in ecological farming, and democratizing this practice is as vital as learning how to farm itself. I've known Dan Brisebois for many years, and he is truly a master at what he does—passionate, highly informed, and rigorous in his approach. His expertise, combined with his innate ability to teach, makes him not only a great mentor but also a remarkable human being. This book is an absolute must-read for anyone who is committed to making a real difference in the world. As you embark on your seed-saving journey, Dan's guidance will empower you to reclaim and preserve the biodiversity that our planet so desperately needs.

—Jean-Martin Fortier, organic farmer and teacher, Market Gardener Institute and author, *The Market Gardener*

I love this book! If you are considering producing seeds on your vegetable or flower farm, or even starting your own seed company, Dan walks you gently and honestly through the weeds with a reminder to take small steps until you find your footing on your seed journey. I look forward to sharing this resource with our new seed growers, staff, and apprentices, and I've already taken many notes for our next growing season. Thank you Dan!

—Owen Taylor, Truelove Seeds

In this practical, down-to-earth book, Dan Brisebois draws from decades of experience to document every step of the seed-saving process, from planning and growing to cleaning and sharing the harvest. *The Seed Farmer* is a fantastic resource for small-scale growers looking to add value to their existing farm business and also empowers home gardeners to take the first step on their seed-saving j

—Erin Benzakein, owner, Floret Farm and au
Floret Farm's A Year in Flowers

This is the book that seed growers have been waiting for! Written in an engaging style that reads like a series of lessons from a mentor, *The Seed Farmer* is full of advice and valuable insights that both novice and experienced seed farmers will find valuable. Crop planning, rough yield data, marketing strategies, and photos of harvest and post-harvest processing are some of the details that earn this book a spot in every seed grower's collection.

—Sarah Kleeger, founder, Adaptive Seeds

Having worked in the seed industry, I can say this book is a must-read for anyone who wants to get started saving their own seeds. Whether you want to grow seeds on your market farm or for your own garden, for your own use or for sale, start here. There are steps that need to be taken for seeds to germinate well and grow true-to type, and Dan Brisebois covers all of them in his down-to-earth writing style with plenty of examples from his own experience.

—Andrew Mefferd, editor, *Growing for Market* magazine

The Seed Farmer is the definitive guide to seed saving for market growers. I'm already putting ideas from it into practice on our micro farm. It's my favorite kind of farming book—practical, inspirational, accessible. Saving seeds is essential work—we can't lose this knowledge. I recommend this book to anyone who grows food.

—Ben Hartman, author, *The Lean Micro Farm*

We are fortunate to receive this distillation of experiential seed wisdom from seasoned farmer and seedsman Dan Brisebois. There are entire universes of subtle nuance to learn and explore within every species of the horticultural panoply. Dan has been delving into this for many years with an exactitude possessed by few others so this tome truly is a welcome Rosetta Stone for seed saving that should find a home on both aspiring and experienced seed savers' library shelves.

—Don Tipping, Siskiyou Seeds and Seven Seeds Farm

Dan Brisebois unlocks something that I don't think most growers even realize is locked: that seed production is possible, even for a busy farmer. Moreover, he presents the "how-to" gently in a way that not only appreciates how complicated seed saving can feel, and how daunting, but gives you an invitation to just enjoy the process. I think that the word "invitation" is perfect for this book. For me, I felt I didn't really know enough about seed saving to confidently fit it into my production systems. But after reading *The Seed Farmer*, I can see it now. I feel permitted to play. I feel invited to the party—directions and all.

—Jesse Frost, author, *The Living Soil Handbook*

This book is a revelation. Somehow, all at once, it's a technical guide, a romance novel, an anarchist manifesto, a masterclass in crop planning, and a botanical encyclopedia, but packed with more inspiration than you'll find in any self-help aisle. It'll take hobby growers to the next level and, even more impressively, I suspect that it will reignite the sense of wonder that many of us flower and veggie farmers have lost over the years. Dan might just make seed farmers out of all of us!

—Lennie Larkin, author, *Flower Farming for Profit*

The Seed Farmer is the only guide I know that is specifically written for the market-scale grower who wants to integrate seed production into their farming operation.

—From the Foreword by Frank Morton, Wild Garden Seed

The Seed Farmer

The Seed Farmer

A COMPLETE GUIDE to GROWING, USING, and SELLING Your Own Seeds

DAN BRISEBOIS

Cover design by Diane McIntosh.

Cover image: © iStock—main image #1744582225 (KharLee); seeds illustration—#1262297025 (m.malinika)
Table image: © iStock-1452497958; Plant vector art: © iStock-2149299152
Interior photos: All images provided by author except where noted.

Printed in Canada. First printing November, 2024.

Any other inquiries can be directed by mail to:
New Society Publishers
P.O. Box 189, Gabriola Island, BC V0R 1X0, Canada
(250) 247-9737

LIBRARY AND ARCHIVES CANADA CATALOGUING IN PUBLICATION
Title: The seed farmer : a complete guide to growing, using, and selling your own seeds / Dan Brisebois.
Names: Brisebois, Dan, 1977- author.
Description: Includes bibliographical references and index.
Identifiers: Canadiana (print) 20240441613 | Canadiana (ebook) 20240441621 | ISBN 9780865719965 (softcover) | ISBN 9781550927894 (PDF) | ISBN 9781771423854 (EPUB)
Subjects: LCSH: Seed technology. | LCSH: Organic farming. | LCSH: Farms, Small.
Classification: LCC SB117 .B75 2024 | DDC 631.5/21—dc23

Funded by the Government of Canada | Financé par le gouvernement du Canada | Canada

New Society Publishers' mission is to publish books that contribute in fundamental ways to building an ecologically sustainable and just society, and to do so with the least possible impact on the environment, in a manner that models this vision.

Contents

Acknowledgments

THANK YOU TO MY TOURNE-SOL CO-FARMERS Emily Board, Frédéric Thériault, Reid Allaway, Renée Primeau, Sophie Descôteaux, Julien Vedel, Camélia Ménard, Calla Sonerson, Amiya Seligman, and Habib Mourad, in addition to all the folks who've been part of Tourne-Sol over the years. I am lucky to farm with such a fun and loving group of people. All your hard work and effort has laid a strong foundation for our seed work.

Thank you to my early seed mentors who helped fill in all the holes that were left after I read all the seed books I could find. Frank Morton, you first inspired me with your seed catalog full of farm-bred originals, and then you took the time to walk me through your fields and drive home the differences between selfers and crossers. Tom Stearns, you've always been there to answer questions about anything from germ tests to seed packing to what songs should go on my next mixtape. And John Navazio, you always insisted on the importance of large population sizes to preserve genetic diversity and potential in a seed variety (even though I tell readers to not worry about that in this book).

Thank you to Andrea Berry, Greta Kryger, and Kim Delaney who started ECOSGN (The Eastern Canadian Organic Seed Growers Network) with me. I miss all the meetups and farm visits we had.

Thank you to all the other seed farmers that I've exchanged with and learned from. I love being part of such an open and welcoming community.

This book wouldn't be here if it wasn't for Julia Shanks and Rob West. Julia, when I told you that I was thinking about maybe writing another book, you immediately set things in motion with New Society Publishers and next thing I knew I was talking to Rob. And Rob, your persistence that I submit a proposal turned into this book. Thank you so much to both of you.

Some of the material in this book first appeared in French as part of the *Guide sur la production de semences à la ferme* which is available on the open source WikiMaraîcher website. This guide was made possible thanks to feedback from

Marie-Claude Comeau, Richard Favreau, Richard Williams, and the support of Hugo Martorell from the Bauta Family Initiative on Canadian Seed Security, a program of SeedChange.

And, speaking of the Bauta Family Initiative on Canadian Seed Security, thank you to Jane Rabinowicz, Helen Jensen, Aabir Dey, Steph Hughes, and all the rest of the folks I've crossed paths with through this initiative. I've learned so much through all the events you've organized and I am really impressed by everything you've done through this program.

Thank you to Aabir Dey, Bob Wildfong, Hugo Martorell, Jared Zystro, Jean-Martin Fortier, Kristine Swaren, Molly Brisebois, and Rebecca Ivanoff who read early versions of this manuscript. Your comments and enthusiasm have made this a much better book.

Thank you to all of you who've been reading the Going To Seed blog and my almost weekly emails. Your support and feedback has motivated me to put the energy into this longer work. And thank you to all of you who have picked up this book and to all of you who share it with your communities of growers.

Thank you to my parents who got me started on this farming path even though I had no idea that that was where I was heading. And thank you to my kids who show me the instinctive wonder at opening bean pods or finding the seeds on a radish plant. You inspire me that we can all be seed growers.

I also want to thank the tobacco plants I found in Lorenz Epinger's fields and the arugula plants at Alison Hackney's farm. If it wasn't for you, I might have lived a very different farming life.

And thank you to Emily Board, my first reader, my co-farmer, and the person who encourages me to keep writing. You have all the best ideas.

Foreword

LOOKING BACK on my 40-odd years of seed growing, I can only say that I wish I had this book to get me started. Learning only through your own mistakes can be a painful process—with *The Seed Farmer*, Dan Brisebois writes with candor and humor, saving "first-seed" explorers a lot of time and trouble. Dan has made plenty of mistakes on your behalf, so that you can get on with making better mistakes of your own. If you want to be a seed farmer, you really want to have this book well-worn by the time you start to make a plan.

The Seed Farmer is the only guide I know that is specifically written for the market-scale grower who wants to integrate seed production into their farming operation. This is exactly the way my own seed career began, when I realized that seed growing was not just a step toward self-reliance, but a positive influence on farm ecology (flowers!), that creates opportunities for economic synergies along with new market possibilities for farm products. It is also true that new *varieties* of vegetables and flowers can be a marketing trait of distinction for both a produce farm and a seed business, and that on-farm plant breeding can be a tool of farming innovation. *The Seed Farmer* takes the reader into the country of seeds to explore the challenges of being a seed producer and a food grower in tandem. But Dan doesn't just take you there, he challenges your thinking with the reality of the work and an economic analysis of whether this will be worthwhile for the farmer—you—to pursue.

This book is written by a highly organized, yet free-spirited, successful market farmer. It is a patient guide into the maze of considerations that go into producing a seed crop while also growing the veg crops that are paying the bills. Dan is a teacher by nature, and his success depends on keeping the first-year seed grower focused on the success of a single seed crop, while keeping the farm in tune. This is difficult, because the slow rhythm of seed cycles is completely different from the weekly rhythm of veg cycles, and while the seed beat is slow, the timing must be perfect, and must always consider the weather.

Dan's teaching insists that the prospective seed grower maintain control of their ambitions for the first year, to take on what he calls a *First Seed Mindset.* He realizes that we all want to run ahead—that we are impatient to get to our imagined futures; but there are important steps to be learned, and techniques to practice, and they are best ingrained by managing 1–3 crops rather than a dozen. There are so many lessons: Label everything. Twice! Keep the harvested plants orderly and parallel, with heads up and butts down! The only two tools necessary for harvesting are a serrated sickle and a tarp. Some seed crops clearly make or save you money, while others are more trouble than they are worth. Cabbage is not a good beginner's seed crop; arugula is.

As we progress from the carefree innocence of *First Seed Mindset,* through the sweet satisfactions of seed self-reliance, and toward the more serious considerations of seed stewardship and selling seed to others, Dan's instruction becomes more detailed and analytical. If you are going to do this work as a business, you need to know the business, and you need to have plans. To have plans you need to be able to make predictions based on calculation. Here, our teacher is a master, showing the planning process and its myriad complications that must account for field rotations, species isolation, seed yields and selling potential, seasonal workloads, crop cleaning or handling challenges, seed dormancy, germination and disease testing... a list too long to consider all at once. He reveals a great principle—much of our planning ahead for what we do not know is working backwards from what we *do* know, and all such calculations must include what he describes as a "safety factor," a wiser term than "fudge factor." This is another way of saying that it never hurts to aim a little high, like 30 percent. Many of us do this in our farm planning, but I really appreciate that Dan includes it as a factor of 1.3 in his calculations. It shows realistic genius.

I like most Dan's instructions to spend time with our plants. He tells us that we need a deep relationship with the crops we steward to seed. We should know them from seeding, through every life challenge, and back to seed again. We must take time to be with them, in all seasons and all weather, to take them in through all our senses, and through all the uses we make of them. This is the way of seed stewardship.

Frank Morton
Wild Garden Seed

You Can Grow Seed on Your Farm

IT'S EASIER THAN YOU THINK

A COUPLE OF GENERATIONS AGO, all of this would have been obvious to farmers, but these days growing seed is a very separate enterprise from growing vegetables and cut flowers. Let me tell you about how I realized it didn't have to be this way when I accidentally wound up with my first big seed harvest.

In the summer of 2002, I was working on a small farm located in the West Island of Montreal. One day while I was doing the rounds tending to crops that we were no longer harvesting, I came upon an arugula planting that had been forgotten in a far corner of the field. And when I say forgotten, these arugula were way past any value for salad greens. These plants were four-foot-tall bushes with brown seed pods and no leaves. I got off the tractor and squeezed a few of these pods. They shattered easily and released orange, red, and brown seeds. I could see that this whole planting was packed with mature seed.

I wondered whether I could collect some of the seed. I cut the plants at the base with secateurs and stuffed them in some old feed sacks. I then stashed the feed sacks in the equipment shed and forgot about them.

At the end of season, we were cleaning up the farm and shutting things down for the winter. Alison, the farmer I was working for, handed me the arugula sacks: "I think these are yours."

I brought my feed sacks of brittle arugula plants home to the kitchen and decided to see what I could do with them. I took handfuls of broken branches and pods out of the bag. The pods seemed empty. I figured the heavy seed had sunk to the bottom of the bag. The top three-quarters of the bag's contents wound up in the compost pile.

Then I emptied the feed sack into a couple of salad bowls. I twirled the bowl in a circular motion. The ensuing whirlpool brought the chaff and remaining pods to the top. I skimmed these off in handfuls to reveal a mass of seeds beneath. This was probably clean enough to reuse on the farm but the seeds still had a lot of chaff mixed in. I wanted these seeds to look like they came out of a new envelope.

At the time, I didn't think to use a box fan to winnow out the light bits. Instead, I poured a pile of seeds onto a baking sheet. I tilted the sheet to the left—the seeds rolled to the left leaving some of the chaff behind. I wiped the chaff away. Then—tilt to the right. The seed rolled to the right. I wiped more chaff away. I repeated the process over and over. The seed got cleaner and cleaner.

My big harvest was about a pound of arugula seed. At the time this was more seed than I knew what to do with. I gave some to friends, I traded some with other seed savers, and three years later, when I started farming at Tourne-Sol, this was the arugula seed we used.

Twenty years later, I'm still working with the descendants of that seed at Tourne-Sol Co-operative Farm. This is the arugula we put in our weekly vegetable baskets and this is the arugula we put in seed packs and sell through our online store.

This was not a complicated process. And I bet it would work well on your farm.

Here are a few lessons hidden in that arugula story:

First, there was no extra bed preparation or weeding for this seed crop. All that work was done anyway for the vegetable harvest. So these bed-feet had already paid for themselves and the work to grow them. (A bed-foot is a one-foot-long slice in your growing bed.) Though leaving arugula plants in the ground to go to seed means you can't use the same space for a second market crop, the financial value of that seed crop might be comparable to a second market crop, especially with those reduced expenses.

Second, even though I had limited seed experience, the arugula that grew from the seed I harvested was as good as the seed we bought from our usual seed company. Growing your own seed doesn't mean compromising on seed or variety quality. In fact, as you'll see later in the book, if you start taking the time to choose the best plants to go to seed, the subsequent crops you grow from that seed might even outperform your starting varieties.

Third, there wasn't much of an investment in equipment to get a seed crop. I did it with stuff I already had in the kitchen. If you were going to invest in some basic seed equipment that would be maybe $40 for two box fans, and five dollars of miscellaneous colanders and spaghetti strainers from the second-hand store. Add in a few tarps you might already have kicking around, and you're up and running.

Fourth, you can pack up the seeds in kraft envelopes with simple printed or handwritten labels to sell at your farmers market during your spring plant sale. You can make money from your seeds without adding an online farm store or launching your own seed company. If you have enough seed you could sell it to another seed company!

Fifth, letting leafy greens go to seed means that there are more flowers in your garden. Those flowers are an ecological blessing to market gardens. They provide nectar to pollinators and predator insects. These flowers can also distract some pest insects like tarnished plant bugs from your other crops. Not to mention that you get to smell the delicious sweet smells of brassicas in bloom.

And sixth, seed work is profoundly satisfying. Turning that mess of feed sacks and broken-up plants into a pile of clean seed is the most amazing magic trick you can do. Each time you clean seed, it gets easier and more efficient but it is no less satisfying. If anything, growing bigger buckets of seed makes that magic more spectacular. Plus you get to experience that feeling of pushing your hands deep into an ocean of seeds and feel them rolling over your skin and between your fingers. And when that seed germinates the next year, the magic is truly complete.

All that to say, growing seed on your farm is not as hard as you might think.

About Tourne-Sol Co-operative Farm

Farming is deeply intertwined with the specific location where it takes place so I want to tell you a bit about where I farm.

Tourne-Sol Co-operative Farm is situated in Les Cèdres, Quebec. We are 45 to 60 minutes west of downtown Montreal depending on road closures and construction.

Our farm is in zone 5a on the Canadian hardiness map and 4 on USDA hardiness maps. We typically can get in the field around May with the last spring frost usually in mid-May, with the occasional early June frost. Summers in our region range from 82°F to 90°F (28°C to 32°C) with high humidity and possibilities of rain throughout the season. Our first fall frost can arrive anywhere from September 19 to mid-October. We're usually out of the field by early November and we get a lot of snow during the winter.

We grow certified organic vegetables, cut flowers, and seeds, using tractor-based farming systems. Our vegetables go to a 500-member CSA veggie basket program that runs from the end of May until mid-November. Our seeds are distributed through an online store and through a seed-rack program. We also grow flowers for a summer bouquet subscription.

Tourne-Sol is run as a workers' co-operative, a type of business where the worker members own the business and share in decision-making and profits. We started with five members in the summer of 2005 and we've grown to ten members and ten non-members, totalling twenty people who farm together.

Throughout this book, I will tell some stories from my seed work at Tourne-Sol to illustrate how the information applies to a real farm and provide you with insights into one farm's journey to integrate seed production into a vegetable farm.

Your story will be different based on your climate in addition to your farming practices, preferences, and challenges. I always appreciate knowing more about the farmers behind the books I read and I imagine you do too!

Some of the 2022 Tourne-Sol team. It's hard to get everyone in one place for a picture!

Credit: Erika Rosenbaum

HOW TO READ THIS BOOK

This book is written with vegetable and flower growers in mind. While I generally use the word farmer, the content of this book is just as applicable to home gardeners.

Whether you're a novice or a seasoned seed grower, this book accommodates individuals of all seed experience levels. If you're new to seed work, you should start at the beginning and read the book sequentially.

If this is not your first seed rodeo, feel free to navigate to sections of interest. However, keep in mind that the book is structured to build upon itself, so starting from the beginning can provide valuable insights regardless of your expertise level.

In Part 1 of this book, we'll dig into the mindset to grow your first seed crop without worrying about all the seed rules. Your only goal in Part 1 is to harvest mature seed.

Chapter 1 guides you through how to choose your first seed crop. It lays out the core botanical concepts that can enrich your first seed growing but also invites you to ignore these rules in order to experiment and learn from your own observations.

Chapter 2 explores the similarities and differences between growing seed crops and vegetables or flowers for market. It highlights the challenges you will likely encounter when you grow both in your fields, and offers management strategies to better integrate seed production into your market garden.

Chapter 3 walks you through your first seed harvest and how to start using your own seed on your farm. This chapter is all about celebrating your first harvests while being cognizant that you are not yet at a point where you can totally rely on the seed you've grown.

In Part 2, you'll learn about what seed crops are a good fit for your farm.

Chapters 4 through 7 cover four types of crops: fruiting vegetables, cut flowers, leafy greens, and root crops. These crops are presented from the easiest crops to integrate in your market farm workflow to the trickiest, with detailed insight into the growing, harvesting, and cleaning particularities of each.

In Part 3, we'll graduate from a First Seed Mindset and figure out how to improve all aspects of your seed work so that you are confident in the seed you grow.

Chapter 8 delves into all the technical skills and tools to handle your seeds from harvest through storage. This chapter also tackles germination tests for your seeds and quite a bit about labelling.

Chapter 9 shifts the focus to a longer-term relationship with seed. Here we'll explore how to make sure you have seed that will outlast you and be the foundation for future farmers. This chapter covers trialling varieties to adapting them to your farm and even breeding new varieties.

Chapter 10 acts as a comprehensive guide to craft a crop plan specifically tailored for seed production. We'll venture into the realm of spreadsheets to explore how to seamlessly integrate the book's insights into planning your upcoming growing season.

Chapter 11 covers the sales opportunities that come with seed. Whether you're looking to sell seed in bulk to other seed companies or contemplating starting your own seed company, this chapter offers invaluable insights and guidance to navigate the market effectively.

By the end of this book, you'll be equipped with the knowledge and skills to grow some amazing seeds! Whether you're a beginner or a seasoned pro, I'm excited to accompany you on your seed-growing journey.

Let's get started!

Part 1
Grow Your First Seed Crop

In the first part of this book I invite you to hold a First Seed Mindset.

I want you to forget about how important high quality seeds are to the success of your vegetable or cut flower farm. Push away your fears and worries that any seed you grow might get cross-pollinated by the wrong plants, or that they won't germinate, or that it's too complicated to understand.

I want you to come to seed with a beginner's mind and embrace everything you don't know and all the experience ahead of you.

Is that a little too trippy?

What is important when you start growing seed is not the actual seed you're going to harvest and later sow to grow crops that you will bring to market. (And it is not about preserving biodiversity or establishing food sovereignty.) Instead the important thing is the work skills that you begin to develop as you start observing and working with plants in this new way.

What will ultimately let you grow good seed and have an impact on your food systems and seed systems is getting better at observing plants, identifying the right moment to harvest, figuring out how to effectively and efficiently handle all that plant matter to extract clean seeds. These are the skills you are honing through each seed crop you experience. And that is what I want you to focus on as you begin this journey.

This is what I call a First Seed Mindset.

Over the next three chapters I will guide you through the basic seed concepts that will have a big impact on how you choose, grow, and use your first seed crops. You might think of these as the rules to the seed game.

You have permission to ignore as many of these seed rules as you want when you grow a seed crop for the first time. This might mean some of your crops don't perform as well as you'd like. But you'll be amazed how many seed crops work out well when you ignore most of the rules.

This advice is obviously for new seed people, but all of us experienced seed growers can also embrace this advice every time we grow a new crop that we have never grown to seed before.

So let's choose what crops to let go to seed …

Chapter 1

Choose Your First Seed Crops

WHAT SEED CROPS should you grow?

The simple answer is that you should grow seeds for the crops you love to grow, from the crops that are already in your garden. These are the crops that you're already building a relationship with and that you depend on.

I have two stories to inspire you with your first seed crops. The first is a short story that involves a walk in the snow and how there are seeds all around us. The second is the story your annual seed order can tell you, if you know how to look at it right, about what vegetables and flowers and herbs would be good seed crops.

Through these two stories you're going to discover that there are quite a few crops from which you could grow seed but I'm going to ask you to choose to grow only one to three of these crops for seed in your first year.

A SHORT STORY ABOUT UNEXPECTED SEED

I spent the 2001 growing season at Switch Farm in Milton, Ontario. It was my second season working on a farm and everything was still new to me. During that season I visited the nearby Everdale Farm (an incubator/apprenticeship farm) a couple of times. While I was there I spent some time with two apprentices: Andrea and Tanya. They were obsessed with seeds.

In the spring, they had found some Red Russian kale plants that had survived the winter and had convinced the Everdale farmers not to till them under so that they could go to flower and then to seed. When I was there, they were stomping on the dry plants spread out on a sheet. I didn't pay attention to most of the steps but I remember being impressed by the jars of clean seed at the end of the day. Until then I had never thought about where the seeds came from on the farms where I worked.

That January, I came back to Milton for a few days to attend the Guelph Organic Conference. I stayed at Switch Farm and took a morning walk through the deep snow to look at the fields. One of my stops was a flower garden we had planted with all the unsold plants from the spring seedling sales. It was a tangle of brown dried-up plants that hadn't been tilled under in the fall.

The tallest plants were tobacco plants. They were taller than me and at their top were little brown berries. I knew that tobacco was in the *Solanaceae* family, just like tomatoes and peppers. I guessed these were the tobacco equivalent of tomatoes. What did the insides look like?

I picked a tobacco berry and squashed it. It was full of dusty stuff. But channeling Andrea and Tanya it occurred to me that these were seeds.

I reached out and plucked five or six of these tobacco berries and put them in my pocket then carried on with my walk. I've been growing tobacco plants from those seeds ever since!

Now this is one way you can get into seeds—simply look at what has gone to seed around you and harvest the seeds. This is a time-honored tradition and it works. And that is essentially what I did with my first big arugula seed harvest from the intro of this book. If you grow cut flowers, you've probably had this happen to you too.

THE STORY YOUR SEED ORDER CAN TELL YOU

Your annual seed order is the list of everything you need for the coming growing season. It is usually the last step of your crop-planning process after you've figured out how much of each crop you'll need to grow for your farmers markets and vegetable baskets and wholesale orders.

If you've already made a crop plan for the coming growing season and you have put together a seed order, then start with that. If you haven't finished this year's crop plan, then go and use last year's seed order. And if you have never compiled your seed order into one sheet, now is the perfect opportunity to do that.

For the following exercise you can adapt your current seed order spreadsheet with the columns I'll mention below; or you can use the Seed Farmer Seed Order spreadsheet at sheets.seedfarmerbook.com

The Seed Farmer Seed Order spreadsheet is a little too big too fit snugly on one page of this book so I've broken it into two tables. Table 1.1 is what your seed order probably looks like. Table 1.2 are the extra columns you can add to your seed order to guide you through what crops would be a good fit for your farm. You can find most of the information for these extra columns in the Crop Profiles in Part 2 and in Appendix 3.

We will now go through columns I through O and consider what they tell you about growing each of your varieties for seed.

Table 1.1

A	B	C	D	E	F	G	H	I
Crop	Variety	Source	Product Code	Qty	Unit Qty	Unit Format	Unit Price	Annual Cost
Bean	Gold Rush	Seed co 3	EX3710	2	5	lbs	$40.00	$80.00
Beet	Touchstone Gold	Seed co 2	EX7086	1	100000	sds	$535.00	$535.00
Carrot	Dolciva	Seed co 2	EX2685	1	500000	sds	$335.00	$335.00
Carrot	Napoli	Seed co 2	EX4275	1	500000	sds	$750.00	$750.00
Cucumber	Corinto	Seed co 3	EX3414	1	1000	sds	$364.63	$364.63
Hot pepper	Early Jalapeno	Seed co 3	EX2633	1	1/16	oz	$8.00	$8.00
Lettuce	Green Frilly	Seed co 2	EX6282	1	50000	sds	$350.00	$350.00
Pepper	Carmen	Seed co 4	EX2954	1	1000	sds	$98.00	$98.00
Radish	Raxe	Seed co 2	EX8767	3	1	lbs	$70.00	$210.00
Tomato	Jaune Flammée	Seed co 4	EX7798	1	2	g	$15.00	$15.00
Tomato	Estiva	Seed co 3	EX8798	2	1000	sds	$41.04	$82.08
Turnip	Purple Top	Seed co 3	EX6634	1	2	oz	$4.60	$4.60
Turnip	Hakurei	Seed co 4	EX2983	1	1	lbs	$225.00	$225.00

- Col A: the crop name
- Col B: the variety of that crop; there is a row for each different variety
- Col C: the source where you got this seed. From a seed company, another farmer/grower, or maybe seed from your farm!
- Col D: the product code or SKU that the seed source uses for that variety. This makes it much easier to communicate with your seed supplier in the future so they know exactly what you're talking about.
- Col E: the quantity of seed packs you're getting of that variety.
- Col F: the numeric quantity of seed count or seed weight in each seed pack. This is measured by the units in Col G.
- Col G: the units in which the seed pack is measured. It might be by seed count or by a weight format such as g, oz, lbs, kg.
- Col H: the price of that seed pack.
- Col I: the total cost for that variety. This is Col E (qty) x Col H (unit price).

Table 1.2

A	B	J	K	L	M	N	O
Crop	Variety	OP/ F1	Certified Organic	Species	Selfer / Crosser	Life Cycle	Type of seed protection
Bean	Gold Rush	OP	Yes	*Phaseolus vulgari*	Selfer	Tender Annual	Protected Dry Seeds
Beet	Touchstone Gold	OP	Yes	*Beta vulgaris*	Crosser	Replanted Biennial	Protected Dry Seeds
Carrot	Dolciva	OP	Yes	*Daucus carota*	Crosser	Replanted Biennial	Naked Dry Seeds
Carrot	Napoli	F1	Yes	*Daucus carota*	Crosser	Replanted Biennial	Naked Dry Seeds
Cucumber	Corinto	F1	Yes	*Cucumis sativus*	Crosser	Tender Annual	Protected Wet Seeds
Hot pepper	Early Jalapeno	OP	Yes	*Capsicum annuum*	Intermediate Selfer	Tender Annual	Protected Wet Seeds
Lettuce	Green Frilly	OP	No	*Lactuca sativa*	Selfer	Hardy Annual	Naked Dry Seeds
Pepper	Carmen	F1	Yes	*Capsicum annuum*	Intermediate Selfer	Tender Annual	Protected Wet Seeds
Radish	Raxe	OP	Yes	*Raphanus sativus*	Crosser	Hardy Annual	Protected Dry Seeds
Tomato	Jaune Flammée	OP	Yes	*Solanum lycopersicum*	Selfer	Tender Annual	Protected Wet Seeds
Tomato	Estiva	F1	No	*Solanum lycopersicum*	Selfer	Tender Annual	Protected Wet Seeds
Turnip	Purple Top	OP	Yes	*Brassica rapa*	Crosser	Replanted Biennial	Protected Dry Seeds
Turnip	Hakurei	F1	No	*Brassica rapa*	Crosser	Replanted Biennial	Protected Dry Seeds

- Col J: whether the seed is open-pollinated or a hybrid.
- Col K: whether the seed is certified organic or not.
- Col L: the crop species.
- Col M: whether the crop is mainly self-pollinated or cross-pollinated.
- Col N: the plant's life cycle.
- Col O: the way the seed is protected on the plant.

CONSIDER HOW MUCH YOU SPEND ON EACH SEED VARIETY

Sort your seed order spreadsheet by column I (Annual Cost) in descending order. This will place the varieties that cost the most at the top of your spreadsheet. The varieties at the top of the list are there because you either use a lot of seed for that crop or the cost per seed is quite high. In either case, growing these seeds yourself on your farm will reduce your seed bill. And in some cases this can be significant, especially when you consider that seed can be stored for many years.

Now I'll remind you that I opened Part 1 by saying your first seed crop is not about good seed, or preserving biodiversity, or selecting that perfectly adapted variety that thrives on your farm better than any commercially bought seed. I will add here that your first seed crop is not about saving a lot of money.

That being said, you might as well learn to keep seeds for some crops that will eventually save money on your farm.

Table 1.3

A	B	I
Crop	Variety	Annual Cost
Carrot	Napoli	$750.00
Beet	Touchstone Gold	$535.00
Cucumber	Corinto	$364.63
Lettuce	Green Frilly	$350.00
Carrot	Dolciva	$335.00
Turnip	Hakurei	$225.00
Radish	Raxe	$210.00
Pepper	Carmen	$98.00
Tomato	Estiva	$82.08
Bean	Gold Rush	$80.00
Tomato	Jaune Flammée	$15.00
Hot pepper	Early Jalapeno	$8.00
Turnip	Purple Top	$4.60

In this seed order, it looks especially tempting to grow your own Napoli or Touchstone Gold seeds. If you grew seed for three years, you'd be saving 3 x $750 = $2,250 for Napoli and 3 x $535 = $1,605 for Touchstone Gold. But it isn't as easy as that!

CONSIDER WHAT VARIETIES ARE OPEN-POLLINATED OR HYBRID

In column J of your seed order, indicate whether the varieties on your seed order are OP or F1. The most important botanical concept for producing seeds is the difference between open-pollinated and hybrid seeds. Then you can filter out the F1s to only show OPs. This is what you can see in table 1.4.

Open-pollinated seeds (also abbreviated to OP seeds) are stable varieties that will give you the same variety the next year as long as you control any unwanted cross-pollination. You should start with OP varieties when you begin growing seed since they are likely to produce what you expect.

Hybrid seeds are created by crossing two plant lines and are marked on seed packs and in seed catalogs with an F1 in their name. The seed you sow will be very uniform but if you keep seeds from a hybrid and grow them out you will get a large diversity of different characteristics. This can be really exciting if you want to develop new varieties but it is very frustrating if you want to reliably grow a specific

Table 1.4

A	B	I	J
Crop	Variety	Annual Cost	OP/F1
Beet	Touchstone Gold	$535.00	OP
Lettuce	Green Frilly	$350.00	OP
Carrot	Dolciva	$335.00	OP
Radish	Raxe	$210.00	OP
Bean	Gold Rush	$80.00	OP
Tomato	Jaune Flammée	$15.00	OP
Hot pepper	Early Jalapeno	$8.00	OP
Turnip	Purple Top	$4.60	OP

Once you've filtered out F1 varieties, you can see that Napoli carrots no longer appear. However, Touchstone Gold beets are still a good seed candidate.

crop that looks a certain way. Do not use hybrid seed varieties for your first seed crops.

What crops are OP vs. hybrid?

Some crops are pretty much only available as OPs. These include beans, lettuce, and peas.

For some crops, most of the market varieties available commercially are now F1 hybrids. These include broccoli, carrots, and onions.

And then there are some crops in the middle where you can find a lot of OPs and a lot of hybrids, like tomatoes, cucumbers, and squash.

CONSIDER WHICH VARIETIES ARE CERTIFIED ORGANIC

In column K, you can indicate whether the varieties you're ordering are certified organic or not. If your seeds are certified organic, this will be indicated by the seed company where you ordered them.

If you are growing certified organic crops, your certifier expects you to be using organic seeds. In cases where organic seed is not available, you will need to provide affidavits from your seed suppliers that they do not offer organic seed for those varieties.

Growing non-organic seeds in an organic system produces certified organic seeds. This is one way for you to develop an organic seed supply of some crops on your farm and should make your organic certifier happy.

CONSIDER WHAT CROPS WILL CROSS-POLLINATE TOGETHER

In column L of your seed order, you can indicate the crop species for each of your varieties. This is the scientific name of a crop. It is often in italics near the top of a crop description in a seed catalog. Generally speaking, two crops of the same species can cross-pollinate and two crops of different species cannot cross-pollinate.

What is cross-pollination?

Cross-pollination occurs when the pollen of one variety is used to pollinate the flowers of a different variety and this produces viable seed. That seed is a genetic combination of both varieties, and when you grow it out you will see characteristics from both parents in the same plant.

If you're looking for a uniform stable variety in your market garden, you might be disappointed with this crossed-up plant.

The key to knowing a crop's species is by learning their scientific name. Crops with different scientific names shouldn't cross together. Beets are the species *Beta vulgaris* and they will not cross with carrots which are *Daucus carota*.

Once you start exploring scientific names you might be surprised to learn what crops are related. Beets and Swiss chard are both *Beta vulgaris*. As such they can cross-pollinate each other. Even though the vegetable part of beets and chard look different, if you let these crops flower, you'll see they are really similar one to the other at that stage.

There are two crop families that really confuse folks: Brassicas and Cucurbits.

Know your brassica species

There are a number of different brassica species. The varieties of different species will not cross with each other. Here are some of the main species you'll find in your garden:

- ***Brassica oleracea:*** Broccoli, gai lan, kale, collard, cabbage, cauliflower, kohlrabi, Brussels sprouts
- ***Brassica napus:*** Rutabaga, Russian kale
- ***Brassica rapa:*** Bok choy, tatsoi, mizuna, Tokyo Bekana, rapini, choy sum, turnip, Napa cabbage
- ***Brassica juncea:*** Spicy mustard
- ***Eruca sativa:*** Arugula
- ***Raphanus sativus:*** Radish, winter radish

So you could grow gai lan, rapini, mustard, and arugula side by side and harvest the seeds and see no cross pollination in the next generation. But if you grow tatsoi and mizuna side by side to seed, you would get some crossed-up greens the next year!

Know your cucurbit species

These are the most common cucurbit species in your garden:

- ***Cucurbita pepo:*** Delicata, spaghetti, acorn, zucchini, pie pumpkin, jack o'lantern, patty pan, crookneck, sweet dumpling
- ***Cucurbita maxima:*** Hubbard, turban, kabocha, buttercup, marrow, banana squash, giant pumpkin
- ***Cucurbita moschata:*** Butternut, cheese pumpkin, musqué de Provence
- ***Cucurbita argyrosperma:*** Cushaw
- ***Cucumis sativus:*** Cucumber
- ***Cucumis melo:*** Melon
- ***Citrullus lanatus:*** Watermelon

You could grow a delicata, a hubbard, and a butternut side by side and harvest the seed and see no crossing in the next generation.

Though two varieties of the same species can potentially cross, you might still be able to grow two or more varieties of the same species depending on how likely they are to cross.

CONSIDER HOW LIKELY CROPS ARE TO CROSS WITH EACH OTHER

Just because two crops can cross-pollinate each other doesn't mean they will cross each other up. And that's what you put in column M of your seed order. There are a number of different ways that plant flowers grow and that pollination happens but for this book we're going to split those ways into three groups based on how likely the crop is to cross-pollinate: Selfers, Intermediate Selfers, and Crossers.

Selfers

These crops barely cross-pollinate at all. Selfers have hermaphroditic flowers that are wrapped in tight petals that make it hard for insects to get to the pollen, and even harder for them to spread that pollen to other plants. This means selfer plants are predominantly pollinated by themselves. (You can probably see why we call them selfers.)

These crops are great candidates for market gardeners who don't have the time or space to isolate plant varieties.

If you have two selfer varieties of the same crop growing side by side and you save the seed of one of them, you will only see one to five percent crossing in the next generation. Now that might be too much crossing for a seed company that wants to deliver true-to-type seeds to their clients. But that's an acceptable amount of crossing for on-farm use.

For your first selfer seed crops, don't worry about isolating selfers from each other—that little amount of crossing won't impact your market garden that much.

Intermediate Selfers

These crops also have hermaphroditic flowers and mostly self-pollinate but you will see higher rates of crossing when you grow varieties of the same species side by side (in the 1–15 percent range). That still might be an acceptable level of emerging diversity in your field. Increasing isolation distance to 100–300 feet between varieties of the same species should greatly reduce crossing.

In each crop profile (in Part 2), I'll talk a little bit more about pollination considerations for these crops. That being said, I don't think you should worry too much about isolating intermediate selfers either for your first seed crops. (Except for keeping sweet peppers away from hot peppers.)

Crossers

Crossers are on the other side of the spectrum from selfers. These are plants that predominantly cross-pollinate. Some are hermaphrodites and some have distinct pollen-producing and pollen-receiving flowers. Whatever the flower structures, crossers want to cross. If you grow two crosser varieties of the same species side by side, you will see 20–50 percent crossing, maybe more. That is too much crossing if you want anything predictable in your market garden.

You might think that this makes these crops hard for market gardens to grow to seed. But for most vegetable crossers that actually isn't the case. For most crossers market gardeners harvest the vegetative part of the crop for their vegetable baskets or market stalls. At that stage they aren't in flower and don't risk crossing up with your seed crop. If you only let one variety of a crop go to flower, there will be no other variety to cross-pollinate your crop. This makes it easy to harvest seeds that will produce plants that are similar to their parents. (In a few cases you might need to be aware of wild weedy relatives that can also cross with your crop.)

That means you might have beds of salad greens such as Tokyo Bekana, mizuna, and tatsoi in your fields. Those crops are all *Brassica rapa* and have the potential to cross with each other.

But you could still safely grow a *Brassica rapa* seed crop in close proximity to your salad beds as long as you keep those salad beds from going to flower.

For crops that are harvested as cut flowers or as fruit crops (cucumbers, squash), you will have more challenges controlling pollination if you grow a lot of diversity.

If you do want to save seed from more than one crosser crop in a year, you will need to grow these crops with at least 1,000 feet between your varieties. This will almost completely reduce crossing to near zero. (The exception being if you have neighbors who grow large fields of commercial seed crops; in those situations there can be a lot of pollen in the air that makes some crossers difficult to grow.)

For your first crosser seed crops only grow one variety of a crop species to reduce cross-pollination complications.

Table 1.5

Selfer	Intermediate Selfers	Crossers
Lettuce, Beans, Peas,Escarole, Poppies, some Tomatoes	Peppers, Eggplants, Okra, other Tomatoes	Pretty much all the other vegetables and flowers.

CONSIDER YOUR CROP'S LIFE CYCLE

Column N of your seed order is where you indicate your crop's life cycle. This is when you need to plant your crop to be able to reliably get it to seed. There are three types of plant life cycles for commonly-grown vegetables: annuals, biennials and perennial. Annuals are usually the easiest to grow to seed.

Is Cross-Pollination All That Bad For Your Seed Crops?

On one hand there are a few reasons why you might want to avoid cross-pollination:

1. You might lose the important culinary traits for a crop—it will become too fibrous, lose sweetness, not have the right moisture content.
2. Your clients might want a specific look or appearance for a crop. If you don't meet their expectations, they might just move on to the next stall.
3. You could affect traits that are important for market growers such as their days to maturity or ease of harvest.
4. You are stewarding a unique variety with a specific story and you want to keep that story intact.

These are all good reasons why you might want to keep a variety stable.

On the other hand, even if a crop crosses up:

1. It will very likely be edible—and in many cases still be delicious.
2. It might exhibit characteristics that you have never seen before—and this might even give you an edge in your marketplace.
3. This can be the beginning of a great breeding project (much more about that in Chapter 9).
4. You'll have less pressure to keep it from crossing in the future because it has already happened!

In fact, there are some growers who really push for growing crossed-up populations of plants. Joseph Lofthouse refers to these as landraces in his book *Landrace Gardening* and promotes growing crossed-up populations as a way to adapt crops to the areas you are farming. Joseph Lofthouse farms in Paradise, Utah, in an arid climate at a high elevation, and writes that he is only able to mature watermelon and other crops by growing these crossed-up landraces.

Practically speaking, as long as you only have low rates of crossing in your crops, you don't have to worry much. If you see any accidental crosses, simply skip those plants when you next harvest for seed. Only keep seed from plants that look the way you want the variety to look.

However, if there is something crossed-up that seems interesting, you might want to keep that seed. But be warned, when you grow it out you will see a full range of diversity and it can get tricky to know what to do with all that diversity. That's the topic for Chapter 9. It's also the chance to expand your mind about what is possible and go down a plant breeding rabbit hole that changes your whole farm.

Annuals

An **annual** is a plant that grows vegetative parts and then goes to flower and sets seed—all in the same growing season. Annuals are usually the easiest crop life cycle to work with. You plant the crop at the beginning of the season and just let the plants do their thing and then you get seeds. There are two groups of annuals:

Hardy annuals can tolerate some frost and can be planted as early as your soil can be worked once very deep freezes have passed. Lettuce, peas, and brassica greens are all hardy annuals.

Tender annuals can't tolerate frost and should be planted once the chance of frost has passed. You can also plant these under row cover, mini-tunnels or in caterpillar tunnels to extend your growing season. Tomatoes, peppers, beans, and cucurbits are all tender annuals.

The goal when growing annuals for seed is to plant them as early as they can possibly survive in the season so that they have a long enough season to go to seed. If you plant annuals too late, you risk running into very wet weather or very cold weather at the end of the season. Those conditions can ruin a seed crop.

In some cases, crops will also switch to vegetative growth as the day length gets short at the end of the summer, and you wind up with bolted leafy plants with no mature seeds.

Biennials

A **biennial** is a plant that grows vegetative parts in one growing season and then those vegetative parts overwinter and go to flower and set seed in the next growing season. Because biennials need to get through the winter, biennials can have a few more steps to get to seed than annuals do. These extra steps depend on your climate.

In cold Northern climates such as where we farm at Tourne-Sol, most biennials will not survive the winter if they are left outside without protection.

If you grow in a climate where the winter does get cold but not quite as extreme, biennials will often happily overwinter in the field—in these climates, biennials can be pretty easy to grow.

If you live in a hot climate where the weather doesn't usually get close to freezing, you've got a different set of challenges. Most biennials need a certain amount of cold before they will switch to flowering and seed production. This is called vernalization. In this kind of climate, you would need to bring some biennials into cold storage to simulate winter and trick these plants into flowering. This makes biennials trickier to grow in hot climates. Annuals are so much easier.

There are two groups of biennials:

Root biennials form a storage root and can be treated similarly to other root crops in cold storage. These are the easiest biennials to plant and replant. Root biennials include carrots, beets, rutabagas, onions, celeriac.

Plant biennials grow an above-ground plant that is much bulkier than just a root. These are easiest to grow in climates where they can be left outside through the winter. If you live in a climate where you can't do this, these crops can be more complex to overwinter and don't make for great first seed crops. Plant biennials include celery, chard, kale, collards, cabbage, and other heading brassicas.

Some hardy annuals like brassica greens or spinach can also be treated like plant biennials and grown through the winter.

Perennials

A **perennial** is a plant that keeps living beyond one or two seasons. It might produce seeds in the first year or it might produce seeds after a few years or it might not produce seeds at all for you. A lot of perennials are propagated in ways other than by seed.

CONSIDER WHAT SEED GROWS BETTER IN YOUR CLIMATE

In column O of your seed order, you can indicate the type of seed protection that each crop provides as it grows its seeds. This seed protection has a big impact on how well different crops will grow in your climate.

Technically you can probably grow almost any crop to seed on your farm, but your climate is more favorable for some crops then others. Focusing on the seeds that do best in your climate will increase your seed-growing success and make you much more inclined to keep working with seeds!

Start by considering the amount of precipitation you get in a season. (Temperature also has an impact on how easily seeds can be fertilized and set. John Navazio has a great chapter about this in his book *The Organic Seed Grower.*) Are you in a dry or humid climate? Rain and ambient humidity can cause havoc on a seed crop. They can result in diseases and seeds germinating on the plant. If you farm in a dry climate, you can harvest high quality seed from just about any crop. If you farm in a wet climate, then you need to carefully choose the seed you will save or grow some seed crops in a tunnel or greenhouse.

There are three broad categories for how seed crops fare in different humidity levels.

Naked Dry Seeds grow unprotected on a plant and are completely exposed to the weather. Cilantro and dill are examples you may have seen in your garden. Plants with naked seeds are very susceptible to diseases or to germinating on the plant before they are harvested. They are not a great fit for wet climates. Other naked dry seeds include spinach, beets, chards, lettuce, carrots, and onions.

Protected Dry Seeds grow in pods or in fruits so are not exposed to the weather. Though these often prefer drier climates, they can do well in wet climates. Protected dry seeds include brassicas of all types, beans, peas, and corn.

Protected Wet Seeds also grow in fruits. They are not exposed to the weather. These crops thrive in humid climates. Protected wet seeds include tomatoes, peppers, squash, melons, and cucumbers.

What type of seed protection should you prioritize for your first seed crops?

You should focus on working with crops that are more adapted to your humidity. This way you will likely have weather working with you instead of against you. You can come back to the other crops on your list when you have mastered the crops suited to your environment and you are ready to take the next step.

What We Grow At Tourne-Sol

At Tourne-Sol farm, we have a warm spring and a hot summer. We also have lots of moisture. Moisture that comes as rain, rain, and more rain; even in drought years, the ambient humidity will often be close to 100 percent.

We have the most success with protected dry-seeded and wet-seeded crops. Specifically crops that can set seed in a warm spring and tolerate hot summers and crops that can set seed in hot summers.

These are the crops that we focus on: tomatoes, peppers, eggplants, squash (summer and winter), cucumbers, melons, watermelons, *Brassica rapa, Brassica juncea,* arugula, radishes, kale, beans, and peas.

We also grow a fair amount of lettuce seed in unheated greenhouse areas. Their yield and quality have really increased compared to when we used to grow seed in the open field.

We do grow more than just these crops, but these are the crops for which we dedicate the most growing space.

Your crop mix will depend on your climate and farm infrastructure and will probably look different than ours.

CHOOSE ONE TO THREE VARIETIES TO GROW TO SEED

Now that you've looked at the varieties on your seed order through all these different lenses, it's time to decide which of these varieties you will plan on growing to seed this year.

Choose from one to three varieties. This number of summer seed projects is small enough to not be too much of a burden when you try to fit them into your growing season.

Going through these steps, there are probably already a few crops that are awfully tempting to you. And you can choose whatever crops you want for your farm. If you're still not 100 percent sure, here are my recommendations and one thing you probably shouldn't start with.

My UNrecommendation

First off, do not choose squash as your first seed crops, simply because you're likely growing many different squash varieties of the same species on your farm to sell at market or put in your CSA baskets. All these vegetable squash flowers will happily cross-up with your seed crop.

It is possible to control most of that crossing by growing your squash 1,200 feet apart or by manually pollinating the flowers yourself. But these are extra steps I don't recommend to new seed farmers who also have packed to-do lists.

Put squash aside as one of your first seed crops.

Recommendation 1

For a first seed crop, annuals are definitely easier to grow. If you're only growing one first seed crop, make it an annual. The easiest crop for most farmers to choose is a pepper or a tomato.

You can grow these with your market crops without any isolation and simply keep a few ripe fruit in peak season to extract the seed. These are the easiest crops to start with.

If you're going to commit to two or three seed crops then here are three more suggestions.

Recommendation 2

Grow a salad green—either a lettuce or a *Brassica* green like arugula, mizuna, or Tokyo Bekana. Grow this crop as part of your market crop. Harvest it as a mixed salad green. After you finish harvesting for salad, let the rest of the crop go to seed. You could leave ten bed feet to go to seed if you're feeling conservative. If you're feeling ambitious, let the whole bed go to seed!

Recommendation 3

If you grow cut flowers, then choose a cut flower you know already goes to seed by the end of your season. One of those flowers that even if you deadhead them, you still wind up with mature seed on plants. Calendulas and amaranths are great choices. Crops that you harvest for ornamental pods are also good choices. They just have to wait a bit longer on the plant and they have mature seed. Nigellas and poppies are examples of these.

Recommendation 4

Biennial seed crops are a bit trickier but the challenges of growing biennials, especially root biennials, are not insurmountable. If you're going to grow two or three first-seed crops, then adding a biennial gives you more learning opportunities as you figure out how to get them through the winter. They also give you an extra sense of accomplishment when you get to seed harvest.

If you grow somewhere where the winter is mild and many crops will easily overwinter, then kale or collards are great biennial seed crops to try.

If you live in a very cold winter climate, where most vegetables won't survive the winter outdoors, then rutabagas, beets, turnips, or winter radish are a good choice. Start by growing a market crop and then choose the best looking roots to store through the winter in your cold room and then replant to grow seed the next year.

THE FIRST SEED PLEDGE

Now that you've made your seed choice it's time to commit!

I've written a first pledge for you to help you remember that this first seed crop is simply about starting a relationship with some seeds. The following can be done alone or with your farm team, with family members, or with friends. If it's your style, you can also record yourself and broadcast your resolution to the world on social media. (And if you do do that, please tagme!)

Put your hand over your heart and say the following out loud ...

> I , (YOUR NAME), resolve to keep the seeds from these seed varieties (NAME THE VARIETIES) this growing season.
>
> I do not need to worry about isolation distances or population sizes or even selecting the best plants.
>
> My goal is to simply
>
> - let the variety go to flower then seed
> - extract the seeds from the crop
> - get the seeds clean enough
> - dry the seeds for two to three weeks
> - and then store the seeds in a container
>
> (I will not forget to label the container with a variety name and the year.)
>
> Once the seed is in a container, I will place it with the rest of my seeds so that I will see it next year when I check my seed inventory and I will include a bit of this seed in my crop plan.

Chapter 2

Grow Seed Crops in Your Market Garden

NOW THAT YOU'VE GOT a list of your first seed crops, how hard is it going to be for you to add these seed crops to your market garden?

As you can imagine, it depends on the crop. And the big difference is how much growing time and farm manipulation it takes beyond growing the market part of the crop.

To grow a seed crop you first need to grow the vegetable, flower, or herb. And up until that point your seed crop doesn't look any different. You need to plant, weed, row cover, and do all the things you would normally do to get that crop to a marketable phase. So all your current growing skills can be fully used here.

And in some cases, this is almost all you need to do to get to the seed part of the crop. This is the case for tomatoes. If you can grow tomatoes for fruit then you've gotten to the part where the seed is mature. All you need to learn is how to extract the seed and clean. Easy peasy.

But in other cases, getting to a market crop is only half the growing journey. Think about lettuce. Once you've got to the salad or head stage, there are no seeds to be seen on this plant. There are more growing steps to get that crop to the seed stage. And there are more growing challenges that come along with those steps.

Here are the main challenges that you are likely to encounter:

- Seed crops are in the ground longer than other crops and can wind up being in the way of other farm operations.
- Seed crops have more vegetative material (and seed pods!) than many market crops; this provides extra challenges for lingering humidity to spread disease.
- Seed crops can leave seeds behind that become future weed problems.
- Seed work is extra work you are adding at the peak height of your growing season and creates opportunities for burnout.

This chapter is about the strategies to manage these different growing challenges.

HOW TO MANAGE SEED CROPS THAT ARE IN THE FIELD LONG PAST THEIR MARKET STAGE

When some crops don't fit with the timing for your other crops, this can interfere with the efficiency and effectiveness of your farm. This can result in frustration and compromise other operations.

> *At Tourne-Sol, we sow partial beds of cilantro and dill every two weeks from May to August to keep a steady supply of fresh herbs. We do this because these crops are at a perfect leaf harvest state for one to two weeks and then put up flower stalks and become fibrous. That's when we usually till under these crops. Sometimes we let some dill and cilantro keep flowering for the pollinators or for a seed harvest and till everything else in the bed. This leaves little islands of flowering plants that tractor operators try to maneuver around as they sow cover crops or prepare more planting beds.*

If you only have one or two crops like this over a whole season, it might just cause a bit of grumbling but if you have a hodgepodge of lingering seed crops sprinkled through your farm, it might trigger a nervous breakdown.

Here are a series of ways to mitigate this problem.

Trellis seed crops

One of the problems with lingering crops is that not only are they in the way where they are planted but often they have tall stalks that flop over into other beds, getting in the way of doing work in those beds too.

Trellising plants keeps those plants orderly and constrained to their bed. So at least the impact is limited to where they are planted.

Seed crops are pretty light compared to tomatoes or other crops you might already trellis, so your setup can be a bit simpler.

I like rebar stakes for trellising; they can be used for many years—seemingly forever. Five-foot stakes are tall enough for most crops. You can put one stake every ten feet or so, sunk one foot into the ground. Then put a few lines of twine around the crops. If you're dealing with single rows, you can pinch the crops between two lines of twine. If you have multiple rows in a bed, you might put a row of stakes on the outer rows and only string up the outside of the row, effectively penning your plants between the two outer rows.

Hortonova flower netting is an option for plants that aren't too bushy and out of control.

Fig. 2.1: *Rebar stakes holding back some flowering kale in a greenhouse.*

Grow seed crops on the edge of blocks

The worst place for a seed crop to be is in the middle of an empty block. Those crops break up empty space into smaller harder-to-manage areas. If you place that seed crop bed to one edge of your field block then you can keep the remaining empty area as one unit and not have to break up operations.

Move seed crops somewhere else

You can also dig up the annoying seed crops and plant them somewhere else. Many crops are surprisingly resilient to being transplanted to

Fig. 2.2: *You can set up Hortnova flower netting over a seed crop such as these onions and the plants will grow through it.*

Fig. 2.3: *A bed of flowering seed radishes on the edge of a block of market vegetables.*

other places. I've done this many times with lettuces that have just started to bolt and brassica greens too. I'll often remove a fair amount of the foliage to reduce the impact of the move. And I make sure that they get water right away to help them establish.

You can move seed crops to other places where they won't be in the way. This could be with other long-season crops, or a home garden, or even into containers. Moving seed plants into containers that are close to the barn can be an especially easy way when you're dealing with a dozen or two plants.

Grow seed crops with other long-season crops

Lingering seed crops are problematic because they are in the field longer than the same crops in their vegetable phase. But if you look at your other crops, you'll see that some of your crops are in the field just as long as your seed crops.

Here are some other long-season crops you might have in your fields:

- Tomatoes, peppers, eggplants
- Squash, summer squash, cucumbers, melons, watermelons
- Cut flowers
- Woody herbs such as oregano, thyme, sage, and others
- Leeks, onions, celeriac, parsnips

Let's say you have a bed of arugula that you know you will let go to seed after harvest. You could choose to grow that single bed with other long-season crops instead of in your salad green blocks. This means that your salad green harvesters will have to visit multiple areas on some harvest mornings, but that inconvenience only lasts a week or two and in exchange you don't have seed crops in places you don't want them to be.

When I place seed crops with longer-season crops from different crop families, I always look at the longer-term crop rotation for that block to make sure that I'm not creating host crops for pests or disease that might compromise the crops in the coming years.

Plan for dedicated seed blocks

As you add more seed crops to your farm, you can simply plan on field blocks that are mostly seed crops. You might also include a few of the long-season market crops mentioned above to fill up these blocks.

HOW TO REDUCE HUMIDITY IN SEED CROPS

When I planted my first bean seed crops, I sowed them the same way I would grow any other snap bean. I then harvested these plants a few times as snap beans, planning on leaving the remaining pods to go to seed. As I picked these fresh snap beans, I noticed the beginnings of white mold on the plants. By the time the bean pods should be mature and ready for dry harvest, the white molds had spread through the whole planting. I was disappointed and I reflected on what went wrong.

Lingering humidity in seed crops creates the perfect conditions for bacterial and fungal disease to spread through your seed crop, especially if you handle the crops regularly (such as by picking beans or tomatoes). You need to figure out how to either keep your plants dry or how to get them to dry down by themselves quickly.

Use underhead irrigation to keep plants dry

The first thing to do is to stop adding extra humidity to your plants when they are at vulnerable stages. For most crops this is when they start to bolt. Up until then you can use overhead irrigation just like you might for the vegetable portion of the crop. But when that phase is done, it's time to switch irrigation strategies. That might mean moving to drip tape. Or if your climate and soil type permit it, you might cut irrigation at this point.

If you only have overhead irrigation systems, then you should focus on irrigating earlier in the day so that the plants have time to dry out before going into the evening when there is generally more humidity anyway.

Change your spacing to increase ventilation

Increasing the space between plants will improve air flow and dry down leaves quicker. This will keep disease from spreading from plant to plant. To do this you can reduce the number of rows per bed or increase the in-row spacing for your plants. If you grow your market lettuce three rows per bed with 9-inch in-row spacing, you could grow your lettuce on two rows with 12 inches in-row spacing to give the plants more space.

For direct-seeded crops, you can reduce the seeding density. At Tourne-Sol, we sow snap beans at 8–12 seeds per row-foot. For seed crops we reduced this to 4–6 seeds, aiming for 2–3 plants every foot.

For your first seed crops, don't worry too much about spacing. Just see what happens to your crop, and if it doesn't do as well as you hoped and succumbs to disease, next time grow the crop more spaced out!

Trellis plants to increase ventilation

This not only keeps plants nice and orderly and out of the way, but trellising plants keeps them off the wet ground and also improves airflow between plants. Trellising also has the advantage of reducing some wind damage.

You can trellis almost any seed crop, be it lettuce or brassica greens or flowers. However, for your first seed crops, I wouldn't prioritize trellising unless you find the plants are particularly floppy and in the way. Trellising is one more task during the growing season and if you can get away without it, you can use that time in other ways.

Weed crops to improve ventilation

Weeds and unwanted plants in your garden bed not only compete for nutrients and light, but they also trap humidity in the tangle of plant leaves. Staying on top of weeding all the way until seed harvest will be one more way to keep your plants dry and disease free.

Weed-free beds are also much easier to work with when it's time to harvest and clean seeds.

Grow under covered spaces

If you really want to keep your plant leaves and stalks dry, the most effective strategy is to grow seed crops in a tunnel or greenhouse. This can keep all precipitation off your plants and reduce most ambient humidity. For some seed crops this might be your only way to get consistent good seed yields.

At Tourne-Sol we grow all our lettuce seed in covered areas. Our yields are so much higher than in the field, and we have far fewer plants melting on their way to seed.

It all depends on the variety

All this being said, you will also notice that some varieties of a crop are more disease resistant or less disease prone than other varieties. These will be easier to grow to a mature crop and you might be able to get away with minimal extra measures.

To some extent you can also select and adapt your plants to wetter climates, and we'll talk about that in Chapter 9. But the ideas listed here will give you better results than if you don't do anything to reduce humidity.

HOW TO MANAGE SHATTERED SEEDS THAT COULD BECOME WEEDS

After my great first arugula seed harvest (that I mentioned in the introduction) I jumped into growing all the brassica greens we were using to seed our market crops: tatsoi, mizuna, and different mustards as well as arugula were the early crops we used. I discovered that no matter how hard we tried, we couldn't harvest every last brassica seed. There was always 10–30 percent that escaped. I could see them lying there on the ground as I was harvesting just waiting for a little moisture to germinate.

Once you start harvesting your own seed crops, you'll get to see your own shattered seed on the ground after your harvest. And if you're not careful, those seeds will stick around for a while creating weed problems for years. You need to create a strategy that doesn't just rely on your being a diligent weeding machine.

Be careful what you grow after a seed crop

The worst things to plant after a seed crop are tightly spaced direct-seeded crops such as carrots or salad greens. These will be overwhelmed by quick-growing volunteers.

Fig. 2.4: *Last year, there were kale plants in this spot in the greenhouse. These seedlings are from that shattered kale seed.*

Instead you should plan on planting wide-spaced crops like potatoes, squash, cucumbers, or tomatoes. The wide spacing gives you more room to control volunteers with your weeding tools.

Even better is to grow a cover crop after a seed crop. The shattered seed will germinate amidst the cover crop and get mowed along with the cover crop. (See page 190 for more on integrating cover crops into your seed rotation.)

Leave shattered seeds on the surface

It might seem counterintuitive, but when you're finished harvesting a seed crop do not till the soil right away. If you till the soil, you will bury those seeds into the soil and hide them from the reach of your weeding tools. These seeds will keep germinating in your soil for years. Instead, leave those seeds on the surface undisturbed for six to eight weeks.

Leaving seed on the surface exposes them to two things that will greatly reduce the weed seed bank: First, they will germinate and you can mow them or later till them under. Second, mice and beetles love to eat seed treats lying on the soil surface.

Another alternative to tilling seeds in, is to irrigate your plot and spread a tarp or piece of landscape fabric to encourage those shattered seeds to germinate and then get smothered.

Keep an eye out for bonus market crops

If you're growing salad greens or root crops for seed, you might also find that these shattered seeds grow quite well and produce healthy vegetables. Instead of tilling them under, you could actually harvest these as a bonus market crop.

If you are in a year of abundance and you have bumper crops, you probably don't need to hold on to these. But if it's one of those years where yields are tight, pay extra attention to these volunteers.

Shattered seeds give you a sneak peek

No matter what you do with these shattered seeds, make sure to take time to look at them. They give you a glimpse into the quality of your seed crop. Do they germinate quickly and grow well? Do they look like you expect? Shattered seed is your first opportunity to see if there are unexpected crosses.

I discovered that Yukina Savoy and Pink Petiole Mix mustards were the same species when I saw all these unknown brassicas popping up and I deduced what was happening. (This unexpected cross was the foundation of Rainbow Tatsoi—more on that story in Chapter 9.)

HOW TO STAY BALANCED WITH EXTRA SEED WORK

August and September are our peak harvest months for all our vegetable crops. This is a pretty hectic time where it seems To Do lists can never be fully checked off and the weather is hot and there isn't a lot of extra brain space to think of anything beyond the essential.

This is a tough time for a lot of farmers and it is an easy part of the year to suffer from burnout.

And that's when many of your seed crops are going to be ready to harvest.

This can seem like seed work might be that one drop that is going to overfill the bucket.

Here are some strategies to keep you from getting overwhelmed from adding seed work to your farm.

Start small (you might notice that this is my mantra)

"Start Small" is my main advice on most things farming. The reason I recommend only growing one to three seed crops at first is so there isn't that much seed work to get worried about.

Your seed crop choice can even further reduce those tasks. If one of your chosen crops is peppers or tomatoes, then keeping seed means harvesting a few ripe fruit from a few plants. You open those peppers, rub the seeds onto a paper towel, and place them on a plate in your kitchen to dry (always making sure to label the variety). That's a 15-minute task. Not too much to worry about there.

If you're harvesting seed from some lettuce or brassica there are a few more steps. But starting with a planting of 30 plants is a lot less work then starting with 300 plants, especially if you decide to ...

Only do the timely stuff in peak season

There are some seed tasks that cannot wait. If you don't harvest a brassica seed crop when it is ready, the pods might shatter and drop most of their seed. Or it might start raining for days, pounding your plants into the mud. So, harvest is a seed task you shouldn't postpone. However most of the post-harvest seed work can happen later if you store your harvest in a dry area.

In some cases you might need to do a lot of handpicking of individual seed heads. But in many cases you can cut whole plants above the root. In either case, bring your harvest into a greenhouse, an outbuilding, or a barn and spread it out on a piece of landscape fabric. If you're in a really dry area, you might even be able to spread those crops on tarps outside (always keeping an eye on the forecast of course!).

For the first days after harvest, go and turn the plants once or twice a day to release any trapped moisture, but once the leaves start to dry you can ignore your seed harvest for a while.

When things start to slow down you can go back to those seed crops and clean them with a more relaxed headspace.

Always make a label *on* whatever piles of seed and plants you leave lying around (as well as in your farm log). If you're stealing moments here and there, you won't always know when you're coming back to the task. You want to leave it so when you come back, you know exactly what is what and where things are at. And if you have some material left after removing seed, send that to the compost right away. There's nothing more frustrating then threshing a pile of plants to discover it was an already threshed pile that should be in the compost.

Share your seed plan with your team

Right from the start of the season, tell the people on your farm that you are growing seed this year. Tell them what crops you've chosen, why you've chosen those crops, and where they will be grown in the field.

Keeping your team in the loop is always a good policy but it's extra important when you're making a change on your farm. This is how you avoid Unexpected Surprises. Your returning staff are used to doing things a certain way and your seed crops might not follow the same protocols. In the past your team might have been on the lookout for bolted lettuce to pull or over-mature zinnia flowers to deadhead. Now it's important to leave those crops as they are.

Having your team in on the seed plan means they can also hold you accountable to your seed commitment when things get busy. They can put a bit of pressure on you to keep on track when it's time to harvest or time to clean that seed.

Your team may have never wondered where all the seed they use comes from. Bringing them in on your seed plans can also start them on their own seed journey.

Create reminders and responsibilities to get the work done

You can mark your seed crops with flags or flagging tape in the field to create reminders to make sure you don't accidentally till in these bolting crops but also as visual triggers to go see how these crops are doing and check up on how close to harvest they are.

You can add expected seed work dates such as trellising or harvest to your farm calendar schedule as another reminder.

You can also delegate someone on your team to be responsible for monitoring those seed crops and telling you when they think they are ready. This can free up your brain for other things and then you just need to act when it's time.

Scale up slowly and intentionally

As you add more seed crops over the years the work will increase, but if you focus on what worked well, you can keep from overburdening your August.

If ever you scale into a lot of seed, especially if you're selling seed, you can start to have the conversation of what August tasks you might want to phase out in exchange for seed work.

You can only do so much—so you want to be doing what you want to be doing. If you really like seed work, make it a priority.

But the first thing is to start small!

Chapter 3

Your First Seed Harvest

BY CHOOSING TO GROW at least one seed crop and then actually growing that crop, you've taken the first big step to becoming a seed farmer. Now it is time to complete your journey and get that seed off your plants and ultimately back into the ground to grow your next crop. These are the key steps to graduate from your First Seed Mindset and feel confident that you can grow a reliable seed crop.

GET THE SEEDS OFF THE PLANT

How can you tell if your seed is ready to harvest?

Usually your plants will give you clues: bean pods dry down from a green tender pod to a dried leathery pod, tomatoes turn bright red, cucumbers get big and ugly, yellow lettuce flowers turn to little fluffy umbrellas. These visual indicators are how a plant tells you it's ready for you to give it a hand getting its seed into circulation.

The other way to tell if your seed is ready to harvest is to go looking for mature seed on your plant. What does mature seed look like? It looks just like the seed that you started with! If you check your plants regularly you'll see that the seed starts off small and tender, often green and easy to bend, But as the seed matures it will fill out the pod and become hard.

When you find mature seed, it is time to harvest it from the plant by popping open a pod with your fingers, or breaking apart a flower and pinching the seeds out, or smooshing the seed out of a fruit. You will find crop-specific techniques in Part 2 and detailed seed extracting and cleaning techniques in Chapter 8 but first, do what feels like it should work. Don't be afraid to try stuff and do anything that feels instinctive. Folks keep coming up with new ways to harvest and clean seed.

GET YOUR SEED CLEAN ENOUGH

Your seed doesn't have to be perfectly clean for you to start using it. It just needs to be able to flow through your seeder and your fingers without jamming things up. That means that it's ok to have some dirt, or dust, or chaff mixed in with your seed if you haven't been able to easily get it out.

You can get your seed to that level of cleanliness by pouring it through a spaghetti strainer or colander and winnowing the seed in front of a fan. Alternate these steps a couple of times, and you should have something you can use. In Chapter 8, you'll see how to get your seed much cleaner than that. But just know that for now clean enough is good enough. As you get better at cleaning seed you can always go back and clean old seed lots as you wish.

TEST YOUR SEED TO SEE THAT IT GERMINATES

Once your seed has been cleaned enough, you can sow some of it into a seedling tray. Sprinkling the seed without aiming for an exact seed count is fine. Then water and care for this tray just like you would any trays of seed and see what happens. If the seed pops out of the ground then you are ready to use this seed for real. If the seed doesn't germinate or only germinates a bit, you could try again. But be prepared to be amazed how much of your seed will actually germinate.

USE YOUR SEED HARVEST

And now is the most important part of growing seed: sowing that seed and seeing what happens.

The thought of using your own seed to grow your own market crops probably fills you with conflicting emotions: excitement to be part of the cycle of life and getting to see your seed babies become plant babies; and fear that you've grown crossed-up seed that might never germinate.

These are both healthy emotions to be feeling at this point. The way to assuage that feeling of fear is to be careful in how you first use your seed. Treat your first seed harvest the same way you treat any new variety or seed source you want to grow. Start with a trial to see how it compares to what you're already doing.

Grow ten percent of your crop from your seed

The first time you use your own seed, don't plan on more than ten percent of your plants coming from your own seed. This means that 90 percent of your crop will still be from your usual seed varieties and seed sources.

If you're growing a total of 1,000 tomato plants, then grow 50 to 100 plants from your own seed. This amount means that if you have a problem with your own seed then you'll still be able to achieve most of your harvest goals with your standard crops. But this is also enough plants that if they do well you will be able to see the difference.

Make sure to clearly label the plants from your own seed in your field.

Evaluate how your seed performs

Pay attention to your plants! This is good advice across the board on your farm. But make sure you're keeping an eye on the plants from your own seed to see how they are doing as they grow and once you've harvested them.

Here are some of the things you can pay attention to when you're looking at the plants that emerge from your first seed harvest:

- Did they germinate well?
- Do they look like what you expected?
- Do any look very different? If yes, how many individual plants look different? Is it under five percent of the population or more?
- Do they mature at the right time?
- Are they more or less diseased?
- How does the vegetable look?
- How does it taste?
- Is there anything about these plants that make you think "wow"?

Always answer these questions in comparison to your standard varieties growing beside them in the field. If your tomato seed didn't germinate well but none of your purchased seed germinated either, you can look for the problem elsewhere in your nursery. But if only your own seed exhibits a problem, there is a higher chance that it comes from something you did.

I suspect you will probably be amazed how well your seed performs compared to the seed you've purchased. As you start paying more attention to your plants, you might also notice that some of the varieties you already grow are less uniform than you thought.

If your seed doesn't do great, tweak your seed production

The two things that are most likely to disappoint you are whether the seed germinates well and how much the plant is crossed-up. If your seed performs noticeably worse than the seed you purchased, think about what caused the problem and adjust next year's seed crop plan to compensate for the situation.

If a crop was very crossed-up: can you improve isolation distances? Or can you get more physical barriers between two flowering crops of the same species—buildings, greenhouses, tall plants, other flowering plants of a different species?

If the crop didn't germinate, perhaps you harvested the seed too early or extracted it from the plant too early. Leave the seed on the plant longer, and possibly cut whole plants and let them mature in a protected place longer.

You should probably also skip ahead to Part 2 and read up about any hints I have regarding your specific crop!

If your seed does well, plant more of your seed next year

If you're happy with your seed's results, then increase the amount of plants you grow from your own seed next year. Go to 30–50 percent of your plants from your own seed, depending on how confident you feel, and the rest of the plants from bought-in seed. The year after that, you might move to 50–70 percent. The year after that you might be at 100 percent!

So if you keep growing 1,000 tomato plants year after year:

- Year 2—you grow 300 to 500 plants from your own seed.
- Year 3—you grow 500 to 700 plants from your own seed.
- Year 4—you grow all 1,000 plants from your own seed.

After four years of working with your own seed, you should feel pretty confident that you've been doing a good job.

OVER TIME, DIVERSIFY THE SEED CROPS YOU GROW

As you become comfortable with those first seed crops, you can add more seed crops to your crop plan. Maybe more varieties of the same crops (as long as isolation distances don't become a problem.) And you can add new crops from different species.

By starting small and taking your time, there is more chance that seed growing will seem realistic on your farm, and that you'll persist and develop your seed skills. This is how you become a cornerstone in a resilient seed and farming community system.

Now let's get back to our regular scheduled programming and go deeper into seed crops.

Part 2
Seed Crops That Are a Good Fit for Your Farm

Part 2 of this book is all about the seed crops that integrate well on a market farm. I've organized them in four groups based on the part of that plant that is used as a market crop: fruiting vegetables, cut flowers, leaf vegetables, and root vegetables.

These crop groups are presented in order of seed work complexity—from the easiest crops to the most challenging crops for a market grower to get to successful seed without too much additional headache.

Each crop profile begins with some thoughts about growing that crop from seed and a sidebar about how I work with these crops at Tourne-Sol farm. Then each profile is organized in the following sections:

Section	What is in this section	More info about this topic in
OP Availability	How likely you are to be able to find good open-pollinated varieties of this crop.	Chapter 1
Managing Cross- Pollination	How to keep this crop from crossing up with other varieties of the same species.	Chapter 1
Seed Yield	How much seed you will be able to harvest from a certain area grown of the crop.	Chapter 10
Growing The Seed Crop	How to grow this crop from seed to the point where it is ready to be harvested. I will mainly focus on the steps that are different than if you were growing this crop only for market.	Chapters 1 & 2
Seed Harvest	How to get the seeds off these crops. For most crops, the seed will still be in pods, flower heads, or fruits.	Chapter 8
Seed Extraction	How to get the seed out of the pods, flower heads, or fruits. This will leave the seeds in almost useable state but there will still probably be some chaff and bits mixed in.	Chapter 8
Seed Cleaning	How to get the chaff and little bits out of the seed for this crop. The goal is to get seeds to look like they just came out of a new seed pack!	Chapter 8

A WORD ABOUT REFINED-HEADING BRASSICAS

As you go through the coming sections, you will notice that you won't find cabbage, broccoli, cauliflower, and Brussels sprouts here. Is that just because I wasn't sure whether they were flowering vegetables or leafy vegetables? There's a bit of truth to that. But they are also crops that I don't think are great first seed crops for market gardeners.

They are all biennial and require more work to get through the winter than roots or annuals. They are bulky plants that are more work to dig up and handle. They have a lot of vegetative material that needs to make it through the winter.

To add to that, you probably don't use as much seed of these crops as you do of many other crops. That's a lot of seed work for not a lot of return, especially considering that a lot of the standard varieties for these crops are all F1s. There are some good OPs but they don't perform the same way because they aren't being maintained the same way.

If you do want to work with these crops, I salute you and encourage you to do it! But do know that you will have more challenges than with many of the crops listed in the following chapters.

WHAT ABOUT OTHER CROPS?

Other than brassicas, I've included the main crops folks are growing on small-scale market farms in Canada and the United States. If your favorite crops are not here, think about what crop group they would be in, then consider how you can transfer the seed work related to these crops to your favorite crop. (I guess this brings you back to a First Seed Mindset! It's good to always stay curious and creative.)

Chapter 4

Growing Fruiting Vegetables for Seed

FRUITING VEGETABLES are the easiest crops to get to seed. You're already growing the seed-bearing part. The seeds are those bits inside the ripe fruit. (I'm sure you've noticed them before.) Unlike onions or arugula or most other veggie crops, there are next to no extra steps to get your crops to go to flower and produce mature seed.

In most cases, you can grow fruiting crops exactly the same for seed as if you were growing a market crop. The main exception is that some fruiting vegetables are eaten at an immature stage such as cucumbers, zucchini, okra, or snap peas. For this type of crop you will need to leave the fruit on the plant longer for it to develop viable seeds.

TOMATOES

If you're growing any open-pollinated tomatoes, then tomatoes are a great first seed crop. You don't have to do anything extra to grow the crop or even to harvest it. And there is so much seed in most individual tomatoes, you don't even need that big a harvest to have enough seed for your own needs.

Seed Tomatoes at Tourne-Sol

We grow seed tomatoes quite differently from how we grow market tomatoes.

Market tomatoes are in caterpillar tunnels and unheated greenhouses to add heat in the early and late season and extend our harvest window. These protective structures also keep precipitation off the leaves and fruits, reducing the number of second-grade fruit.

We grow seed tomatoes in the field without any cover. We aren't banking on early tomatoes to fill our vegetable baskets and it's ok if the fruit doesn't look perfect.

Most of the market tomatoes we grow are hybrid seed varieties. We buy new seed for these every year, though we do occasionally keep seed from hybrids to develop new varieties. (You can read about the Ruby-Sol tomato in Chapter 9.)

When we harvest tomato seeds from our market tomatoes, we don't worry about isolation distances. I've seen a bit of crossing but only in low levels, and anything crossed up is still edible. When we grow seed tomatoes for sale, we grow these with 200 feet separating varieties.

OP availability

There are a lot of OP tomato varieties out there. If you're into heirlooms then you're probably growing a lot of OPs and you should keep your own seed. That being said, most market varieties are hybrid seeds.

Managing cross-pollination

Tomatoes are selfers. You will see zero to five percent of crossing if you grow them side by side. You can eliminate almost all crossing with 200 feet between varieties.

Most tomatoes are in the *Solanum lycopersicon* species. Currant tomatoes are usually in the *Solanum pimpinellifolium* species. Currant tomatoes and non-currant tomatoes are able to cross with each other.

Seed yield

The amount of seed per fruit makes a big difference on seed yield. Many paste tomatoes contain few seeds per fruit while cherry tomatoes are jam-packed with seed. The number of fruits per plant also makes a big difference. Cherry tomatoes have hundreds of fruit per plant while some big beefsteaks only have a dozen or two.

Seed Crop Bed Length	Imperial Yield	Metric Yield	Seed Yield
100 bed-feet (30 bed m)	1 to 4 lbs	0.5 to 1.75 kg	200K to 600K seeds
1 bed-foot	0.2 to 0.6 oz	5 to 18 g	2K to 6K seeds
1 bed m	0.7 to 2 oz	16 to 59 g	7K to 20K seeds

Growing the seed crop

Grow seed tomatoes exactly the same way you grow market tomatoes. They are tender annuals, so plant seedlings out after your last spring frost date. Avoid overhead watering of plants to reduce the spread of disease.

Seed harvest

You can pick seed tomatoes as soon as they start to change color but seed quality improves as the fruit ripens. If you harvest tomatoes that aren't fully ripe, let them sit in a protected area for a few days so that they keep ripening before extracting the seed from the fruit.

Seed extraction

Getting seed out of your tomatoes is a three-step process.

1. Scoop out the seeds. (You can eat the remaining flesh.)
2. Ferment the seeds in their own juices for two to four days. This dissolves the gelatinous membrane around the seed, improving seed germination. Fermentation also controls some diseases that might be on the surface of the seed.
3. Spread out the clean seed on a tea towel and let them dry for a couple of weeks.

For large seed lots, you can use the wet seed tools listed in Chapter 8.

Fig. 4.1: *Cut your ripe tomatoes in two.*

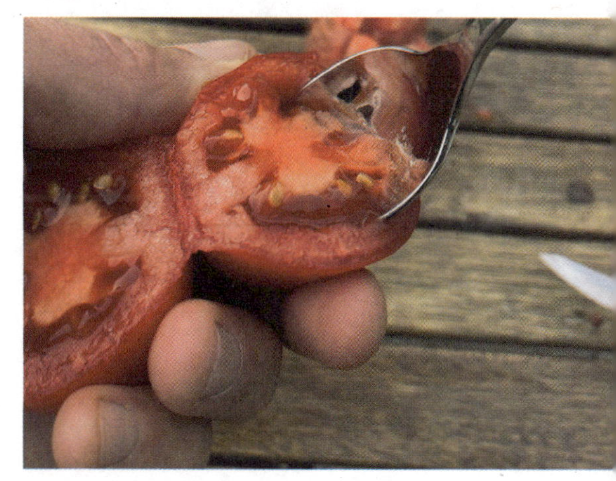

Fig. 4.2: *Scoop out the insides.*

Seed cleaning

Fermenting and decanting the seed usually leaves you with some pretty clean seed. If the seed dries in a clump, you can break it up by rubbing the clump through your hands.

Fig. 4.3: *Drop the scooped seeds and juice into a jar.*

Fig. 4.4: *Stir the clumps to break them up and make sure all the seeds are submerged in the juices.*

Fig. 4.5: *Cover the jar with plastic and poke some holes in it to keep fruit flies out. You can also use a tea towel. Let the jar sit in a warm area. It will start to ferment. Stir/shake the mixture every day.*

Fig. 4.6: *Fill the jar with water. All the light stuff will float to the top.*

Fig. 4.7: *A layer of mold will form on the juice. It's time to decant when the seeds sink to the bottom of the jar. Some tomatoes with very dense juices will not sink. If you add a bit of water and the seed sinks, it's ready to decant.*

Fig. 4.8: *Pour off the mold and everything that floats, stopping just as the seeds get to the lip of the jar opening.*

Fig. 4.9: *You'll wind up with a bunch of seed at the bottom of the jar. Fill the container with water again and repeat the process. Each time you do this the seeds will be cleaner and cleaner. Pour the wet seeds into a colander and spread them out to dry.*

PEPPERS (HOT AND SWEET)

Always remember that hot peppers and sweet peppers can cross-pollinate together. If you have sweet peppers that are accidentally pollinated by a hot pepper, next year you'll wind up with a row of sweet peppers with a couple of plants that have spicy fruit. If you don't catch this mix-up before sending your peppers to market, your clients will figure it out for you when they put them in their kids' school lunches.

Even worse is when hot peppers get accidentally pollinated by sweet peppers. There is no one more disappointed than a chili head who discovers their peppers have no bite. So keep your hot peppers 200 feet apart from your sweet peppers and you should keep the heat where you expect it.

Otherwise, growing pepper seed is even easier than saving tomato seeds! You don't have to do anything extra to grow the crop and a few fruits will produce many seeds.

OP availability

There are a lot of solid OP hot peppers out there. You can probably shift your hot peppers completely over to OPs. Though there are some great OP sweet peppers, a lot of market standards are F1.

Seed Peppers at Tourne-Sol

At Tourne-Sol, we don't have a long enough season to get all our peppers to mature in the field. For market crops, we grow most of our hot peppers in the field, but we grow sweet peppers and some of the longer-season hot peppers in caterpillar tunnels to get them to mature more quickly and extend the harvest season. We only grow OP hot peppers and a mix of F1 and OP sweet peppers. For a long time, we would just keep seed from our market crop without worrying about isolation. We have seen some crossing (more than with tomatoes) but as long as we keep sweet and hot peppers a couple of hundred feet apart we've been fine. I love finding crossed-up peppers and always harvest their seed so I can grow some out the next year to see what they give. (This is how my Carrot Bomb hot pepper started—more in Chapter 9.)

I prefer to grow seed peppers in the field; this lets me gradually select varieties that perform better outside. We still grow longer-season varieties in our unheated greenhouse because we want to get enough seed for sale. I grow seed peppers with 250 feet isolation distance to keep them true to type.

Managing cross-pollination

Most peppers on your farm probably fall into one of two species: *Capsicum annuum* or *Capsicum chinense*. There are a couple of other pepper species out there too. In my experience, it's best to assume that all pepper species can cross with each other, even if they don't have the same species name.

Peppers are intermediate selfers. You will see five percent or so crossing if plants are planted side by side. Two hundred and fifty feet should reduce crossing to almost zero.

If you really want to grow two varieties very close together without crossing, you can cover them with row cover. Since peppers do not need insects for pollination, they can pollinate in this way. Be sure to check plants regularly though—it is easy to lose control of weeds that are hidden this way or to have aphid problems explode on you. There is also less ventilation which can lead to disease conditions.

Seed yield

Seed Crop Bed Length	Imperial Yield	Metric Yield	Seed Yield
100 bed-feet (30 bed m)	1 to 4 lbs	0.5 to 1.75 kg	80K to 240K seeds
1 bed-foot	0.2 to 0.6 oz	5 to 18 g	800 to 2.4K seeds
1 bed m	0.7 to 2 oz	16 to 59 g	2.8K to 8K seeds

Growing the seed crop

As long as your peppers are able to turn color, you can grow seed peppers exactly the same way you grow market peppers. If you're in a climate where you're only able to harvest peppers at a green stage then you should use tunnels or a greenhouse to extend your season by a couple of weeks to get your fruit to mature.

Peppers are tender annuals. Plant the seedlings out after your last spring frost date. Peppers are generally more tolerant than tomatoes to overhead watering; but if you can avoid overhead watering, that would be best.

Seed harvest

Only harvest mature fruit for seed. Generally that is easy to tell for peppers that start green—they mature to a bright yellow, red, or orange. But some immature

Fig. 4.10: *Slice your ripe peppers in half.*

Fig. 4.11: *Reveal a fruit full of seeds.*

Fig. 4.12: *Rub the seeds off the middle of the fruit to get very clean seeds that don't really need further cleaning.*

peppers start purple or light yellow—these will also change color when they are mature.

You can pick peppers as long as they have begun changing color. Let the fruit sit in a protected area for a few days to keep ripening before starting extraction.

Seed extraction

For small batches, slice your peppers open and remove the seeds by hand. Spread the seeds on a plate to dry for a week or two. Use the emptied fruit in your hot sauces or other dishes.

If you're working with hot peppers, be careful! Wear gloves to keep the spiciness off your skin. Wear a mask when you're working with especially hot peppers.

When you're processing large amounts of pepper seed, you can crush the fruit by foot (make sure to clean your rain boots between batches!) and then screen out the flesh. You can use water to separate the lighter bits from the heavier seed. You can also use the wet seed tools listed in Chapter 8.

Seed cleaning

Pepper seeds extracted by hand don't require further cleaning. If you've crushed the peppers you might need to winnow and screen out any lingering bits of dried pepper flesh.

EGGPLANTS

You eat eggplants at an immature fruit stage so you can't just pick an eggplant and extract the seeds. You have to leave the fruit on the plant for another month or so to get the seeds to mature. Still, though it should be as easy to grow eggplants for seed as it is for tomatoes and peppers, eggplant seed has really frustrated me.

Over many years, I have not been able to get consistently high germination from my own eggplant seeds. Some seed lots germinate above 95 percent and some wind up at 40 to 60 percent. Now 40 to 60 percent germination rates might not be a problem for a market gardener—just sow extra seed. But as a commercial seed company, we can't sell that kind of seed.

Some advice I got a couple of years ago was to only leave a couple of fruit to mature per plant. Just pick off the rest and the plant will focus its energy into the few fruit it has left. This also makes sure that those remaining eggplants have an extra-long season to turn big and brown and yellow and full of nice fat seeds packed with germination power. This approach works well for a market garden where you can keep selling all those eggplants you keep picking. Just make sure to leave some first fruits to get nice and ugly.

For commercial seed production, leaving only a couple of fruit per plant can produce a lot of seed for big-fruited eggplants but small-fruited eggplants might not give you much seed. For small-fruited eggplants, you might want to leave more fruit per plant and cross your fingers.

OP availability

A lot of market eggplant varieties are F1 hybrids and are very productive. There are many OP eggplants available but you should trial them to see if they meet your needs.

Managing cross-pollination

Solanum melongena is the most common North American eggplant species. There are also other eggplant species, notably *Solanum macrocarpon*. Theoretically, different eggplant species have the potential to cross but I have not seen it.

Eggplants are intermediate selfers that need more isolation distance than tomatoes or peppers to reduce the risk of crossing. Five hundred feet between varieties

Seed Eggplant at Tourne-Sol

At Tourne-Sol, we mostly use hybrid eggplant seed for CSA. Since F1 seed doesn't breed true to type, this means we have not been keeping much eggplant seed in the market garden. We do grow OP eggplant seed for sale through our seed catalog. We grow these with at least 500 feet of isolation distance and don't see any crossing.

should do the job. For farm use, you can get away with minimal to no crossing and enjoy the occasional off-type.

Similarly to peppers, you can isolate eggplants with row cover for seed production, but the plants are so big it is definitely a hassle.

Seed yield

Seed Crop Bed Length	Imperial Yield	Metric Yield	Seed Yield
100 bed-feet (30 bed m)	1 to 3 lbs	0.5 to 1.25 kg	120K to 300K seeds
1 bed-foot	0.2 to 0.5 oz	5 to 14 g	1.2K to 3K seeds
1 bed m	0.7 to 1.6 oz	16 to 46 g	4.2K to 9.6K seeds

Growing the seed crop

Grow eggplants for seed mostly the same way you would for a market crop. Just leave the fruit on the plants longer than you would a vegetable crop. Flag the first fruit on a number of eggplant plants and don't pick them early! You can pick all the rest of the fruit on the plant without worry.

You should let the fruit mature for four to five weeks before harvesting it. Eggplants turn brown or yellow as they become mature.

Seed harvest

Cut a fruit open to check if the seed is maturing. If there is mature seed in the fruit you should harvest the fruit. You can leave harvested mature eggplants for five to ten days if you don't have time to extract them immediately, but make sure they don't rot away on you.

Seed extraction

For small amounts of eggplant seed, you can easily squish the seed out of the fruit, rinse it with water, and then let it dry.

Extracting large amounts of eggplant seed can be frustrating. Eggplant seed is tightly lodged into the flesh of the fruit and it takes a lot of time to squish it out. One trick is to cut off the top part of the fruit that doesn't have any seed in it to reduce the amount of matter you're handling. Tearing or cutting your eggplants into smaller pieces makes them easier to handle. At this point you squeeze the fruit over and over to remove the seed or use one of the wet seed tools in Chapter 8.

Seed cleaning

Rinsing eggplant seed usually leaves you with some pretty clean seed.

Fig. 4.13: *Cut off the top of the eggplant where there is no seed.*

Fig. 4.14: *Cut two lines at the top of the eggplant—not too deep because you don't want to cut any seed.*

Fig. 4.15: *Using those cut marks as starting points, tear the eggplant into chunks that are easier to handle.*

Fig. 4.16: *If you only want enough seed for yourself, put the pieces into a bucket with water and then squeeze the seeds out of the fruit by hand. Otherwise, run them through a seed extraction tool.*

OKRA

If you can grow okra for market, you can grow okra for seed. Simply leave a few pods on the plants and watch them get big, then dry down to woody pods. In colder climates, you'll need to rely on early maturing varieties to make sure you have enough time.

Seed Okra at Tourne-Sol

Even though there is an okra pod split open on the cover, I have not grown a ton of okra in my farming career (full disclosure: the picture is a stock photo that makes the cover look great). The last few years I have been catching up on lost okra time by growing out larger and larger plantings. We're pretty successful with shorter-season okra varieties and are selecting pods from the earliest plants as seed stock to see if we can get the plants to be even a bit earlier.

OP availability

Though there are some F1 hybrids on the market, there are a lot of OP okras out there.

Managing cross-pollination

Okra is in the *Abelmoschus esculentus* species. It is an intermediate selfer with big showy flowers. There will be some crossing if you grow two okra varieties side by side—500 feet between varieties should eliminate almost all crossing. If you do get crossed-up okra, they will still be edible.

Seed yield

Seed Crop Bed Length	Imperial Yield	Metric Yield	Seed Yield
100 bed-feet (30 bed m)	10 to 30 lbs	4.5 to 13.75 kg	67.2K to 205.8K seeds
1 bed-foot	1.6 to 4.9 oz	45 to 136 g	700 to 2.1K seeds
1 bed m	5.2 to 16.1 oz	148 to 446 g	2.2k to 6.8K seeds

Growing the seed crop

You can grow okra for seed the same way you'd grow it as a vegetable. However, you need to leave the fruit on the plants longer than you would for a vegetable crop. Flag the pods on a number of okra plants and don't pick them. You can pick all the unflagged fruit without worry.

If you grow in a warmer climate, you can harvest quite a few okra pods from a plant before you leave some pods for seed. But northern growers should leave the first pods for seed to make sure they have enough time to mature.

It will take six weeks after the plant sets okra pods for the seed to mature.

Seed harvest

The seed should be ready when the fruits become woody. Open a couple of pods to see if the seed looks mature. Harvest individual pods with secateurs and put them in a bin. If you put these pods in a dry area, you can extract the seed later.

Seed extraction

For smaller quantities of seed, twist each pod between your hands to split it open and release the seeds. You can work the seeds out by hand.

For large quantities of seed, you can try crushing the seed but the pods don't shatter the same way many other seed pods do. You won't be able to release all the seed as effectively this way. You can use a thresher to extract more seed.

Fig. 4.17: *Grab an okra pod in your hands.*

Fig. 4.18: *Twist your hands in opposite directions to split open the pod.*

Fig. 4.19: *Spread the pod open with your thumbs.*

Fig. 4.20: *Separate the parts of the pod to release the seed into a container.*

Seed cleaning

If you extract okra seed by hand it comes out pretty clean and you don't need to do much else to it.

BEANS

Beans can be such an easy crop to grow to seed—they are selfers, they are easy enough to clean, and they are so beautiful to look at. Yet beans are one seed crop I always tell market growers to be wary of considering.

The reason to think twice before growing your own bean seed is that if you pick fresh beans for market then there will be less pods on the plant to dry down with mature seed. Unlike a crop like okra or eggplants where leaving a few dry pods is enough to get all the seed you need, you need a lot more bean seeds than a few bean pods will provide.

Another challenge in wet climates is how easy it is to spread disease between plants when you're picking beans, and if you're spreading white mold through your planting it can easily be covered in rotting pods that will ruin your seed.

This means it is usually best to grow dedicated seed bean plantings for your seed needs. This winds up taking extra space on your farm that you could dedicate to a higher-revenue-generating crop. Beans are often a low-profit crop on small market farms, so having to use even more space for the same crop can really impact the amount of room you have for crops that generate higher sales.

All that being said, you might still choose to grow your own bean seed. Maybe it's the only way you can get the varieties you love. Maybe you also want to harvest some to eat at home. I can relate to all those reasons but if these are your first seed crops on a market garden, you might be dismayed by how much space they take.

Seed Beans at Tourne-Sol

At Tourne-Sol, we grow bean plantings for seed with at least 50 feet distance between varieties. We do not pick the plantings for vegetable harvest to avoid spreading any disease. We've found it hard to grow enough seed for both our farm use and our seed company, so we buy a lot of bean seed from other growers.

OP availability

Almost all bean varieties are OP. You can keep the seeds from your favorite market standards without worry of it not being true to type.

Managing cross-pollination

Most bush beans and pole beans are in the *Phaseolus vulgaris* species. They can all cross together. This is a different species than lima beans (*Phaseolus lunatus*) or runner beans (*Phaseolus coccineus*). Those two other species will generally not cross with *Phaseolus vulgaris*.

Phaseolus vulgaris beans are selfers and you will see very little crossing if you grow your varieties side by side. A few crossed-up plants isn't a big deal considering all the beans will be edible. However, when you grow out your beans, it is pretty easy to miss that you have a few crossed-up plants until harvest time. This can mean some handpicking if you want to remove these beans from next year's seed stock.

If you also grow dry beans, you should keep them 50 feet away from your snap beans since dry beans have different culinary characteristics that you wouldn't want to mingle with your snap beans.

Seed yield

Seed Crop Bed Length	Imperial Yield	Metric Yield	Seed Yield
100 bed-feet (30 bed m)	15 to 20 lbs	6.75 to 9 kg	24K to 33K seeds
1 bed-foot	2.4 to 3.3 oz	68 to 91 g	200 to 300 seeds
1 bed m	7.9 to 10.8 oz	223 to 298 g	800 to 1.1K seeds

Growing the seed crop

Growing beans for a seed crop is similar to growing a market crop. The big difference is that you need to leave the pods on the plants another two months or so for them to dry down. You should consider growing bush bean seed crops with a bit more space between plants than you might for snap beans. This provides more ventilation and can reduce disease. This is especially useful in wet climates.

Theoretically you can pick your bean crop and then leave the last picking to go to seed. However this often doesn't work out for two reasons:

1. Picking fresh pods significantly reduces yields unless you have a very long growing season.
2. Handling plants and moving to the next plants can transmit diseases and infect your whole planting, especially in wet climates.

If you want to grow your own bean seed, I recommend that you cordon off a section of your plot and dedicate it to bean seed. As they say: You can't pick your bean and save it too.

Seed harvest

When the pods begin to turn yellow/tan, you can pick the pods. These can be left to dry for a few weeks before you extract the seed.

Fig. 4.21: *Split your pods in half*

You can also harvest whole plants when they are two-thirds ready and let them mature in a protected space like a greenhouse. If the beans are advanced enough, you can actually pick plants with pretty green pods and still get them to mature this way. Let the plants dry down for a couple of weeks before extracting the seed.

Seed extraction

Hand shelling individual pods is a great way to spend time watching TV, listening to the Seed Growers podcast, or chatting with friends. Just split the pod over a bin, dropping the seeds into the bin, and toss the empty pods into a compost container.

If you harvested full plants, you can strip the pods off the plant and then hand shell them as above.

You can also thresh bean plants by stomping on them, running them over with a tractor, or using a thresher. The drier the weather and the plants, the easier the pods will shatter. However, if it's too dry some beans can also split in half. This can be a bigger problem when using a thresher.

Fig. 4.22: *Use your fingers to slide the beans out of the pods.*

Seed cleaning

Hand-shelled beans don't need much extra cleaning. Threshed beans can be cleaned using the dry seed techniques from Chapter 8.

In the weeks after you've cleaned your bean seed, check to see how dry the beans are by biting on a bean. If you smoosh the bean, it still needs to dry. If instead it cracks then it is dry enough for storage.

Fig. 4.23: *You can stomp on beans plants to shatter the pods.*

Fig. 4.24: *Or run them through a thresher like this modified soil shredder.*

Fig. 4.25: *Once you remove any big stems, you'll be left with some beans mixed with split pods and chaff.*

Fig. 4.26: *Screen your beans through a ½ inch screen to remove the split pods and any big bits. Here's our field screening station.*

Fig. 4.27: (Left) *Pour the screened bins into another bin while the wind blows the chaff away. This is called winnowing.*

Fig. 4.28: (Right) *You'll wind up with mostly clean seed. Winnowing a few more times should get this pretty clean.*

Before storing your beans, you can put your fully dried beans into a freezer for one week. This will get rid of any weevil eggs that might be hiding in your beans. After you remove the beans from freezing, let your containers sit for a few hours so that they get back to room temperature before you open them. This will reduce the chance of unwanted condensation.

PEAS

Peas are just about as easy to grow for seed as beans are. They are also selfers and they dry down well in a variety of climates. They can take a bit more work to clean but not a ton.

And like beans, I caution market growers to think twice before growing these for seed. And for mainly the same reason—you use a fair amount of pea seed to sow a pea planting and harvesting peas as vegetables dramatically reduces the seed yield of your pea plants.

On the positive side, peas don't seem as ready to succumb to white mold as beans so if you did decide to pick the first flush or two from your plants and then keep the rest for seed, you could. You just won't have as good a yield as a dedicated seed pea planting.

OP availability

All peas are OP. You can keep the seeds from your favorite market standards without worry of it not being true to type.

Managing cross-pollination

Snow peas, shelling peas, and sugar snap peas are all in the *Pisum sativum* species. They can all cross with each other.

Peas are selfers and though I usually say don't worry about crossing when growing selfers, that's not true with peas. If you're growing two different types of peas, make sure to isolate them from each other by at least 50 feet. Different types of peas have very different eating characteristics and you don't want them to get mixed up.

Seed Peas at Tourne-Sol

We grow pea plantings for seed with at least 50 feet distance between varieties. Occasionally when a vegetable pea planting is too mature for harvest, we'll let it finish maturing and harvest it as a seed crop. Like beans, we've found it hard to grow enough pea seed for both our farm use and our seed company, so we buy a lot of pea seed from other growers.

Seed yield

Seed Crop Bed Length	Imperial Yield	Metric Yield	Seed Yield
100 bed-feet (30 bed m)	15 to 20 lbs	6.75 to 9 kg	24K to 33K seeds
1 bed-foot	2.4 to 3.3 oz	68 to 91 g	200 to 300 seeds
1 bed m	7.9 to 10.8 oz	223 to 298 g	800 to 1.1K seeds

Growing the seed crop

Growing peas for a seed crop is similar to growing a market crop. It takes five to six extra weeks for pea pods to dry down with mature seed. Peas don't need to be spaced as far apart as beans do for good ventilation in wet climates. You can grow with the same spacing you would for market crops.

Though you could pick your crop and leave the last pods to go to seed, your yield will be much lower than if you leave the plantings unpicked at all. To be sure you get the yield you need, you should cordon off a section of your planting to only be harvested as seed.

Seed harvest

Wait for the pods to turn yellow/tan to pick the pods. You can store dry pods for a few weeks before you extract the seed.

You can also harvest whole plants when two-thirds of the pods mature and bring them into a protected space to keep drying down and mature some of the remaining pods.

Fig. 4.29: (Left) *You can let pea pods and plants dry down in the field.*

Fig. 4.30: (Right) *When you're handling large quantities of peas, you can pull whole plants once they're dried and stack them on pieces of landscape fabric.*

Seed extraction

Like beans, you can choose to hand-shell small amounts of seed by hand; or thresh large amounts of seed by stomping on them, running them over with a tractor, or using a thresher. Dry weather makes it easier to shatter pods.

Seed cleaning

Hand-shelled peas don't need much extra cleaning. Threshed beans can be cleaned using the dry seed techniques from Chapter 8.

Fig. 4.31: *Roll up the landscape fabric and stack the rolls on a wagon or cart.*

Fig. 4.32: *You can leave dry plants in a greenhouse or dry outbuilding until you have time to deal with them.*

Fig. 4.33: *Rolling on dried pea plants with a tractor is one way to thresh your seed crop.*

Fig. 4.34: *Gather the dried threshed material and pour it over a screen.*

Fig. 4.35: *Shake that screen to let the seed fall through.*

Fig. 4.36: *These peas are ready to be winnowed.*

You can also freeze peas using the same method described for beans to protect against weevils.

SQUASH (SUMMER AND WINTER)

Getting seed out of a squash and cleaning it is incredibly easy. Plus you wind up with a cleaned fruit that is ready to get roasted in the oven or chopped into some soup. But don't let that fool you. Squash is a tricky crop to grow for seed in your market garden.

The squash problem is that it has those big showy flowers and is a crosser. There will be a lot of cross-pollination in your fields. Way more crossing than you can ignore. Yes, any crossed-up fruits are generally edible, but they aren't all tasty and some might be excessively stringy. You don't know how good a mystery will be until you bite into one. Unlike peppers or tomatoes or lettuce, selling mystery squash is risky business.

Note that if there is crossing, you won't see it in your squash—it will be in the DNA of the seed you keep. And you might only be able to recognize it the next year when you grow that seed out—and more precisely, when the fruit starts to set.

The other tricky thing about squash is that there are actually four different squash species that behave differently from one another and that usually don't cross with each other. If you grow any two of the same species, you will get some crossing. You need to either limit yourself to one squash of a species for your market garden, or have a separate seed squash garden that is 1,000–1,500 feet from your market squash garden.

There is also the option to hand-pollinate your squash flowers. This is not something I would recommend for your market garden. It is an extra task that will definitely conflict with other priorities in the heat of the season.

Seed Squash at Tourne-Sol

At Tourne-Sol, we grow a lot of squash and zucchini for our vegetable baskets. This makes it a real challenge to also grow squash for seed. For many years, a neighbor let us grow a quarter-acre seed plot in the middle of their organic corn field. We called that plot the Secret Garden because no one knew where it was other than the folks who drove out to work in it.

Every summer we grew a Pepo, a Maxima, and a Moschata squash in the Secret Garden. The Secret Garden didn't have any irrigation, which wasn't usually a problem since most of our early summers are wet. But we did have a couple of very dry summers that hindered germination. And another problem—when your garden is secret, it's not on the top of the weeding list. That's how the Secret Garden becomes the Weed Jungle Garden. Squash definitely has more jungle tolerance than other crops so this only impacted yield a bit.

In the end we got tired of having to go off-site to work in the squash plot so we quit growing our Secret Garden. These days we have other seed growers grow squash seed for us on contract.

OP Availability

Though there are a lot of F1 winter squash and summer squash market standards, there are also many good OPs.

Managing cross-pollination

Squash can be one of four species: *Cucurbita argyrosperma, Cucurbita maxima, Cucurbita moschata,* or *Cucurbita pepo*. If you're only growing one squash variety from each species, you won't see any crossing. It's when you're growing two squash of the same species you need to be careful. Squash is a crosser so you need

Species	Types
Cucurbita argyrosperma	Cushaw
Cucurbita maxima	Hubbard, Turban, Kabocha, Buttercup, Marrow, Banana Squash, Giant Pumpkin
Cucurbita moschata	Butternut, Cheese Pumpkin, Musqué de Provence
Cucurbita pepo	Delicata, Spaghetti, Acorn, Zucchini, Pie Pumpkin, Jack o'Lantern, Patty Pan, Crookneck, Sweet Dumpling

Seed yield

Seed Crop Bed Length	Imperial Yield	Metric Yield	Seed Yield
100 bed-feet (30 bed m)	4 to 11 lbs	1.75 to 5 kg	12K to 36K seeds
1 bed-foot	0.6 to 1.8 oz	18 to 50 g	100 to 400 seeds
1 bed m	2 to 5.9 oz	59 to 164 g	400 to 1.2K seeds

1,000–1,500 feet to see next to no crossing in your squash. If you're growing a lot of squash, you might find that you still get some crossing even with those distances.

Growing the seed crop

You can grow seed winter squash exactly the same way you grow market squash. They are tender annuals, so plant seedlings out after your last spring frost date. Avoid overhead watering of your plants to reduce the spread of disease.

To get summer squash to have mature seed, you need to leave the fruit on the plants for an extra six to eight weeks.

Seed harvest

When your squash has changed color and the stem has dried, the seed will be mature. You can leave the seed in the squash for many months before extracting it. The seed quality will actually improve if you let it mature a while like this.

You can bring squash into your house and extract the seed as you eat them gradually through winter. You can select for taste and storage ability but you should also mix the dry seed with the rest of the same variety to make sure you're growing out more than one individual.

Seed extraction

To get the seed out you need to split the squash in half and then scoop the seeds. For some thin-skinned squash a sharp knife will do the job of splitting them, but for squash with tougher rinds, use a sharp hatchet. For larger amounts of squash you can use a wood splitter.

Fig. 4.37: *Slice open that squash with a hatchet.*

For small amounts of squash, a good spoon can do the job of scooping. For larger amounts, you can use a shop vac to scoop the seeds out. This is so much easier and quicker than scooping away with a spoon. It also separates some of the pulpy fibers from the seed, making it a bit easier to clean the seed later. Petra and Matthew from Fruition Seeds taught me the shop-vac technique and it really speeds up squash seed processing.

If you split your squashes, you can bring the half-squashes home to roast and fill your freezer with more

Fig. 4.38: (Left) *Vacuscoop out the insides with a shopvac to remove the seeds from the fruit.*

Fig. 4.39: (Above) *Empty your vacuscooped seeds from your shopvac.*

bags of squash than you might know what to do with. In theory, you can also sell these hollowed half squashes, but you would have to process these in a food-grade kitchen to meet most health standards. I'm not sure whether vacuscooping squash seed is permitted in a food grade kitchen.

For even larger squash seed lots there are squash extracting machines!

Seed cleaning

The tricky thing with squash seed is that the seeds are wrapped in fibrous strands with small chunks of squash flesh. With small amounts you can pick out the seeds and rinse them off with water.

If all your squash seed sinks, large amounts of seed can be easy to clean by putting the seed in a bucket of water and pouring off what floats. You'll discover that not all good squash seed sinks; some seed lots will have good seeds that float too. This makes cleaning seed with water hard to do. As an alternative, I've found that putting squash through a one-inch screen will remove the biggest chunks of fibers and flesh. Then I spread the squash onto a screen that is smaller than the seeds, and spray the seed with water, aiming to get the fiber to go through the screen, leaving the

Fig. 4.40: (Above) *Rinsing seed coming out of the cucurbit seed extractor at Semences Saint Laurent.* Credit: Marie-Claude Comeau

Fig. 4.41: (Left) *The extractor in action.* Credit: Marie-Claude Comeau

seed on top. When the screen gets clogged, I put the seed in a container and then clean the screen with water or a brush and then put the seed back on and spray some more.

It's okay if some fibers remain on the seed; these can be removed after the seed is dry. Too many fibers though can make it harder to dry your seed.

To dry your seed, spread it out in a thin layer and use fans to blow air through the seed. Stir regularly during the first two days to keep the seed from drying in one big pancake. Once the seed is dry on the outside of the shell, you can pile the seed higher. But you still need to keep drying the seed, preferably with fans. Fresh squash seed has a pretty high moisture content and this can cause it to quickly mold if it is put too early into a container with no air circulation. Ideally you would let your seed dry for two to three weeks before storing it. Waiting longer might even be better!

As the seed dries, it will also shed light fluffy seed wrappers. You can winnow this and any dry fibers out using the dry seed techniques from Chapter 8.

Warning: Some growers will let their squash seed ferment for a few hours to a day or two to help loosen the fibers. This will help speed up seed cleaning but I've found that it can also reduce the seed germination. I do not recommend that you ferment the seed this way.

CUCUMBERS

If you've ever come upon a few forgotten cucumbers that have turned into a gnarly yellow fruit then you are pretty close to a seed crop. Cucumbers are easy to get to seed and they are much easier than squash seed to clean. The biggest challenge to growing your own cucumber seed is if you're growing many types of cucumbers in close proximity. Pickling cucumbers, English cucumbers, Lebanese cucumbers—all these varieties will cross. Crossed-up cucumbers will still be edible but they might not be good for the specific uses that you have in mind for each type of cuke.

Seed Cucumbers at Tourne-Sol

At Tourne-Sol, we used to select our favorite cucumber plants in the field for seed. Most of the cucumbers were slicers and we didn't mind if there was a bit of crossing. Nowadays we grow our seed cucumbers about 1,200 feet away from our market cucumbers. We have limited isolations like this so we only grow one seed crop in a given year. We contract with other seed growers to produce other cucumber varieties for us.

OP availability

There are a lot of good OPs out there but most market standards are now F1 hybrids. These F1s have been bred for all-female plants or to be parthenocarpic (they don't need to be pollinated). If you rely on F1s, you might find that OPs behave a bit differently than you expect. Trialling a number of OPs might let you discover something that meets your needs.

Managing cross-pollination

All the different varieties of cucumbers are in the *Cucumis sativus* species. They are crossers and will happily cross with one other. One thousand feet between two varieties should be enough to avoid most crossing.

There are also Armenian cucumbers and snake cucumber varieties that are in a different species. They are *Cucumis melo*—the same species as cantaloupe and other similar melons. These two cucumbers will not cross with *Cucumis sativus* cucumbers.

Seed yield

Seed Crop Bed Length	Imperial Yield	Metric Yield	Seed Yield
100 bed-feet (30 bed m)	3 to 6 lbs	1.25 to 2.75 kg	50K to 100K seeds
1 bed-foot	0.5 to 1 oz	14 to 27 g	500 to 1K seeds
1 bed m	1.6 to 3.3 oz	46 to 89 g	1K to 3.3K seeds

Growing the seed crop

You can grow seed cucumbers almost exactly the same way you grow market cucumbers. The main difference is that market cucumbers are usually harvested at an immature stage before the seed has developed. Cucumbers are tender annuals, so plant seedlings out after your last spring frost date. Avoid overhead watering of your plants to reduce the spread of disease.

Seed harvest

Leave your cucumbers on the plant for an extra six weeks beyond when you would normally pick them to eat. Your cucumbers will get big and turn yellow/orange when the fruit is fully mature. There might still be quite a bit of green on some cucumber varieties when they get to this stage. You can double check that the seed is ready by cutting the fruit in half and looking at the seed. If the seed has filled out and looks like cucumber seed then it's ready.

Fig. 4.42: *Carefully cut your cucumbers in half without cutting into the middle of the fruit.*

Fig. 4.43: *Look at all those intact mature seeds in the middle of that cucumber.*

Fig. 4.44: *Scoop them out with your fingers, into a bucket.*

You can let the fruits sit for an additional two to three weeks after harvest before extracting the seed. Just keep an eye on your cucumbers to make sure that they don't rot during that time.

Seed extraction

The basic technique is to cut your cucumber in half. Take care to only cut through the flesh of the fruit and not into the middle where the seeds are. Scoop the seeds out into a container, using your fingers. You can use a spoon to do this but I much prefer using my fingers.

Gather the scooped juice and seeds into buckets and let the buckets ferment for up to two days. The mixture will become very frothy and the seeds will sink to the bottom. Pour off the liquid and collect the seeds in a colander with a tight enough mesh that the seeds don't go through. Spray off any remaining froth and any of the sticky bits from the seed. You can then put the seed back in a bucket and fill it with water. The seed will sink to the bottom and you can pour off the floating gunk to leave you with clean seed.

Spread the wet seed out to dry just like you would for squash seed.

Seed cleaning

Cucumber seed is usually quite clean after extraction. but there are often immature seed that might be mixed in. Use the winnowing techniques from Chapter 8 to remove this light seed.

Fig. 4.45: *Let your bucket of cucumber seeds and juice ferment for up to two days.*

Fig. 4.46: *When your cucumber seeds are fermenting they will get nice and frothy.*

Fig. 4.47: *Pour your cucumber seed mixture into a colander and spray off the froth.*

Fig. 4.48: (Above) *Your cucumber seed should be pretty clean. You can decant it once or twice if you like.*

Fig. 4.49: (Left) *Spread your seeds out to dry or put them in a bag and place them in a forced-air dryer.*

MELONS AND WATERMELONS

Melons and watermelons both have "melon" in their names and they are both deliciously sweet and they are both quite easy to get the seeds out of the fruit. But melons and watermelons are two different species. If you grow melons or watermelons on your farm, you might be able to set yourself up for a good seed crop.

OP availability

There are many OP varieties available for both melons and watermelons but this is a crop where most of the market standards have been taken over by F1 hybrids.

Managing cross-pollination

Melons are in the *Cucumis melo* species. Watermelons are in the *Citrullus lanatus* species. The two species will not cross with each other. Armenian and snake cucumbers are also *Cucumis melo* so they will cross with melons.

If you want to grow two varieties of the same species with minimal crossing, you should keep them separated by at least 1,000 feet. However, crossed-up melons will still be tasty and much more pleasant than crossed-up surprise squash.

Seed yield

Seed Crop Bed Length	Imperial Yield	Metric Yield	Seed Yield
100 bed-feet (30 bed m)	1.5 to 2.5 lbs	0.75 to 1.25 kg	30K to 40K seeds
1 bed-foot	0.3 to 0.4 oz	7 to 11 g	300 to 400 seeds
1 bed m	1 to 1.3 oz	23 to 36 g	1lk to 1.3Kseeds

Seed Melons and Watermelons at Tourne-Sol

We can easily grow melon seed at Tourne-Sol. That being said, we have an unreliable season to grow delicious melons on our farm. The heat doesn't always come at the right time and we can have too much or too little moisture. Without being able to consistently get delicious melons, we don't grow them for market.

On the other hand, watermelons can get sweet enough for us, but the season always feels a bit too short for the seeds to fully mature. We grow a few watermelon varieties for market which means we can't grow melons for seed without them cross-pollinating on our farm. We contract watermelon seed from other seed growers who don't also grow fields of market watermelons.

Growing the seed crop

Growing melons and watermelons for seed is identical to growing them for market. They are tender annuals, so plant seedlings out after your last spring frost date. Avoid overhead watering of your plants to reduce the spread of disease.

Seed harvest

The first thing to do is follow your usual indicators for telling how a specific variety is ripe. For melons it might be the smell or whether the first tendril next to the stem is brown or whether the melon slips off the stems easily. For watermelons, it might be if there is an echo when you slap the fruit or if there is a yellow spot on the bottom of the fruit.

Whatever your indicators, the real trick is if you slice the fruit open and find mature seeds inside then the seeds are ready.

If the seeds aren't mostly mature, leave the fruit on the vine longer. You can get some seed ripening once the fruit is removed from the vine but only if the seeds have reached a certain point.

You can put harvested melons and watermelons aside for many weeks before you extract the seed. Just check on them regularly to make sure they haven't started rotting. You may also find that the fruit isn't so tasty if you let it sit for many extra weeks.

Fig. 4.50: *Melons and watermelons are different species and won't cross together.*

Fig. 4.51: *Split your melons and scoop out that seed!*

Seed extraction

For melons, split the fruit in half and scoop out the seeds from the middle. Put them in a bucket of water and let them sink to the bottom then pour off everything that floats. Rinse the seeds. If the half melons are still tasty, eat them!

For watermelons, you can also split them open and remove the flesh and put it in a bucket of water. You can squish everything up and pour off whatever floats. You can do this a few times and get your seed clean. Another option for watermelons is to get your farm crew together for a watermelon break. Have everyone spit out their watermelon seeds and collect them in a container. Then rinse those seeds with water.

Fig. 4.52: *Gather your team together to have a watermelon break. Make sure everyone collects their seed and puts them in the seed bucket.*
Credit: Shaina Hayes

Seed Corn at Tourne-Sol

We don't grow corn for our CSA but we do grow some for our seed company. We alternate between growing a couple of beds of popcorn or sweet corn for seed. We choose early varieties that will flower before the neighboring dairy farm's corn goes to flower. We inspect every harvested ear to make sure there are no crossed-up kernels.

Seed cleaning

Melon and watermelon seed are usually quite clean but there are often some immature seed that might be mixed in. Use the winnowing techniques from Chapter 8 to remove this light seed.

SWEET CORN, POPCORN, AND OTHER TYPES OF CORN

If you can grow corn, then you can grow corn seed. And mature corn seed is quite easy to clean. The biggest challenge with corn is that it is a crosser and there is a lot of corn (for animal feed and silage) in many landscapes that will cross with your corn. It can be hard to avoid getting your corn crossed up.

But are you even growing sweet corn on your market farm? You don't see a lot of sweet corn on small market farms these days. It takes a lot of space for a couple of ears of corn. Sweet corn is more present on larger market farms, but in that case most of the varieties are F1 hybrids.

OP availability

Most commercial sweet corn varieties are F1 hybrids. There are still OPs out there, but most OP sweet corn does not taste like the corn you buy from road stands and doesn't stay at peak tender sweetness for long. Make sure to grow them and taste them before committing heavily to these varieties.

There are many OP popcorn varieties out there.

Managing cross-pollination

There are many types of corn—sweet corn, popcorn, dent, flint, or flour corn. All these different types of corn are in the *Zea mays* species and they will all cross with each other. If you are in an area that is full of corn fields, there will be corn pollen in the air and risks of cross-pollination. What is more problematic is that a lot of that corn is GMO corn—and it will cross with your market crop with abandon.

Generally 1,000 to 1,500 feet will keep crossing to a minimum but if there is GMO corn in your neighborhood, small amounts of crossing is a problem.

Sweet corn has one advantage for seed farmers. You can actually see crossed-up kernels on your seed cob the same year that it gets crossed. Sweet corn kernels are skinny little shriveled kernels. Any kernels that crossed with field corn will be big fat kernels. If you pick these out, you can keep yourself from growing crossed-up corn.

Popcorn has one additional edge: most popcorn varieties will not accept the pollen from other types of corn.

Seed yield

Seed Crop Bed Length	Imperial Yield	Metric Yield	Seed Yield
100 bed-feet (30 bed m)	10 to 15 lbs	4.5 to 6.75 kg	24K to 36K seeds
1 bed-foot	1.6 to 2.4 oz	45 to 68 g	200 to 400 seeds
1 bed m	5.2 to 7.9 oz	148 to 223 g	800 to 1.2K seeds

Yields are for sweet corn. Other types of corn can produce up to twice this yield.

Growing the seed crop

You can grow seed corn the same way you grow corn for market. Sweet corn is not yet mature so you will need to leave the ears on the plant for an extra four to six weeks to mature. The seed on other types of corn is viable when the corn is mature.

Seed harvest

You can start harvesting corn when the kernels have begun to dry down. Snap off the cobs and put them in sacks or bags for transport. Spread them out in a protected area where they will keep drying down. You can remove the husks if you want. You can leave the corn until later to process.

Seed extraction

Pull back the husks from the ear, inspect the ear for any crossed-up kernels and any diseased kernels. If you see a crossed-up kernel remove it.

Once you've inspected your corn you can rub two ears of corn together so the kernels will pop off. You can also purchase corn-shelling rings that help with the job. Or you can run corn through a thresher.

Seed cleaning

Threshed corn can be cleaned using the dry seed techniques from Chapter 8.

Fig. 4.53: *As we husk the corn, we inspect each of the cobs to make sure they don't contain any crossed-up kernels.*

Fig. 4.54: *If we have husked cobs drying in the greenhouse, we'll cover them with screen or hardware cloth to keep rodents from them.*

GROWING OTHER FRUITING VEGETABLES FOR SEED

The first question to ask when you're harvesting a fruiting vegetable for seed is whether the fruit you eat is at an immature stage and if you'll need to leave the fruit longer on the plant. You can check this by opening the fruit and looking to see if there are mature seeds. If the fruit is eaten at an immature stage, it generally needs four to six weeks for the seed to mature on the fruit. Once you start to see well-formed seeds, you can generally harvest the fruit and put it in a ventilated warm area for five to ten days to keep maturing.

There are many other fruiting vegetables you might grow to seed but here are some additional hints for a few crop families:

Solanaceae *fruiting vegetables*

Ground cherries are selfers, tomatillos are intermediate selfers—they are not the same species.

You should assume that most *Solanaceae* crops are intermediate selfers until you've discovered otherwise by trial and error. If you're only growing one variety of a species then you don't need to worry about crossing. Some *Solanaceae* crops can be cleaned quite effectively simply by putting the seeds in a bucket, filling it with water, and then pouring off whatever floats. If you're struggling to free *Solanaceae* seed from the flesh of their fruits, try fermenting them like tomato seeds for a few days—this will usually make them easier to clean.

Fabaceae *fruiting vegetables*

Edamame and soybeans are selfers in the *Glycine max* species.

Black-eyed peas and yard-long beans are both in the *Vigna unguiculata* species so they can cross together. They are selfers though, so a 50-foot distance is probably enough to avoid most crossing. That being said, be careful to keep the two types of crop separate unless you want to have 15-foot-tall black-eyed pea plants.

Many other beans are intermediate selfers (such as limas, runners, and favas) and you should grow them with 500 feet of distance between varieties of the same species. If you're only growing one variety of a species then there won't be any crossing.

You can use similar extraction and cleaning techniques for *Fabaceae* crops as you would beans and peas.

Cucurbitaceae *fruiting vegetables*

This includes bitter melon, luffa, cucamelon, and jelly melon, among others. These are all different species, and if you look at the seeds inside this will be very obvious.

Assume that *Cucurbitaceae* crops are all crossers and if you are growing more than one variety of the same species, aim for at least 1,000 feet isolation distance.

For your first attempts cleaning a new-to-you *Cucurbitaceae* seed crop, start by scooping or squishing out the seeds and seeing if they will sink in water. If so, pouring the floating bits off will be the easiest way to clean these seeds.

If you have a hard time getting the seed clean for *Cucurbitaceae* crops, you can try fermenting them for one or two days.

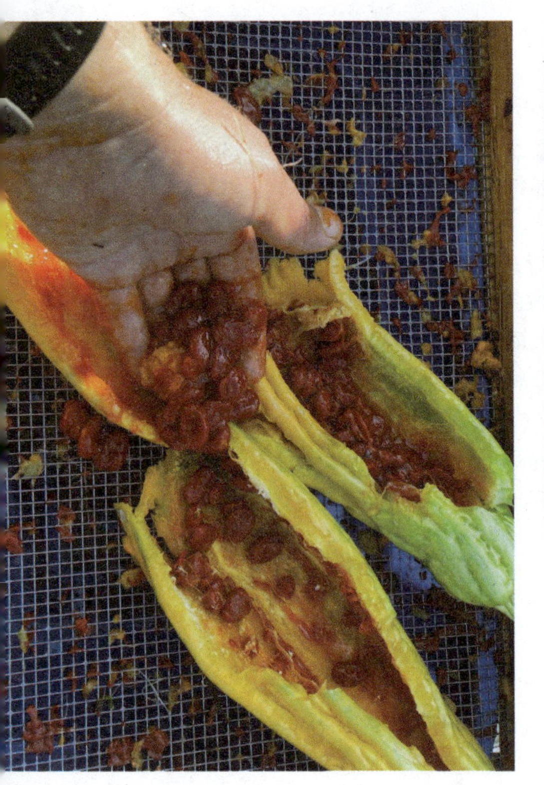

Fig. 4.55: *Mature bitter melon seeds are wrapped in a red membrane. Scoop the seeds out of the fruit.*

Fig. 4.56: *You can spray the red membranes off the seeds with a hose. The brown seeds are mature. The white seeds are not fully mature and will get winnowed out when the seed is dry.*

Chapter 5

Growing Cut Flowers for Seed

CUT FLOWERS are the second-easiest type of market crop to get to seed. You are already growing the crop up to the stage just before seed. Generally at this point you cut the blooms you need for your sales channels and then deadhead the rest. To get seed, you need to not deadhead some of those flowers and let them do what they want to do (which is go to seed).

Don't wait too long to stop deadheading the plants you're going to let go to seed. It can take longer than you think for seeds to fully mature, so give them more time to make sure they finish before your growing season gets too wet.

There isn't as much information about growing cut flower seed as there is vegetable seed. But don't let that stop you from growing the seed and learning as you go.

You should generally assume that cut flowers are crossers. They have those showy petals, after all, that call over all the pollinators in the yard. So if you want minimal crossing give them about 300 to 500 feet between varieties of the same species. If your flower is a crosser, you might still see a five to ten percent crossing at this distance. If you want to avoid almost all crossing, aim for 1,200 feet and more between varieties.

If you're aiming for very specific varieties and color palettes then no crossing is important. But if that isn't the case, you should embrace any crossing you do see. This is your chance to see new shapes and color combinations. You might wind up seeing things that you haven't seen in any seed catalog or flower book. Make sure to harvest seeds from any crossed flowers that you fall in love with—we'll talk in Chapter 9 about what to do with that seed. If you don't fall in love with some of those crossed up blooms, just pick them and put them in your bouquets!

You might also discover that you don't get any crossing at 300 to 500 feet distance. Through experience I've discovered that some flowers (such as poppies) behave much more like selfers. If you suspect a flower is more of a selfer, try growing it with no isolation distance to see if you get any crossing.

Pay attention to the fact that some flower types have multiple species that go under the same flower name. In most cases these different species won't cross together.

Now there are a lot of cut flowers out there and these are crops that I'm still exploring. I'm presenting here those that I have the most experience with to get to seed.

AMARANTHS

Whether you're growing amaranth as a cut flower, a bunching green, or as baby leaves, it is pretty easy to get to seed and harvest way more than you need. If you're growing amaranth for grain then you've already gotten to seed!

OP availability

There are very few F1 hybrid amaranth varieties and a lot of OPs.

Managing cross-pollination

There are many amaranth species and I have a hard time telling them apart. From what I read the most important distinction is between *Amaranthus tricolor* and all the other amaranth species. *Amaranthus tricolor* doesn't cross with other amaranths, but you should assume all the other species might cross amongst themselves until you have experience otherwise. This means you should pay attention to weedy amaranths on your farm.

My experience lends me to believe that amaranth is an intermediate selfer. I have not seen much crossing between amaranth varieties grown 300 feet away in our fields.

Seed Amaranth at Tourne-Sol

We grow a few different amaranths for cut flowers. It seems there are always a few blooms that don't get picked and they quickly go to seed. Self-seeded red amaranths pop up all over the farm as a result.

For a number of years, I tried to cross up a few different varieties of amaranths but there was always less crossing than I hoped. As a consequence, when I grow a dedicated seed amaranth crop, I grow it about 500 feet from our cut flower beds and feel pretty confident that there won't be any crossing.

Seed yield

Seed Crop Bed Length	Imperial Yield	Metric Yield	Seed Yield
100 bed-feet (30 bed m)	5 to 20 lbs	2.25 to 9 kg	2.7M to 11.2M seeds
1 bed-foot	0.8 to 3.3 oz	23 to 91 g	27K to 112K seeds
1 bed m	2.6 to 10.8 oz	75 to 298 g	88K to 367K seeds

Growing the seed crop

Plant your amaranth seed crop as early as possible after your last frost to make sure you have enough time to get to seed. During the vegetable phase of your crop, you can irrigate with sprinklers, but you should avoid any overhead irrigation once the crop starts to go to flower.

It is easier to work with amaranth plants for seed if they are spaced in rows 24 inches apart with 12 inches in-row spacing. But you can get a good seed yield from even tighter spacing than that without too many disease problems.

You can harvest amaranth plants for cut flowers or greens and still get a lot of amaranth seed.

Seed harvest

Keep an eye on the amaranth panicles. At one point you'll start seeing a few black or white seeds sticking out of them. (There are black-seeded and white-/golden-seeded amaranth varieties.)

If you need a few thousand seeds, you can easily harvest that by rubbing seed off a few panicles into a bucket or bin. If you want bulk seed, you can harvest individual branches that have begun to show seeds and bring them indoors to continue drying down. You can also harvest whole plants if you wish.

Spread out stems or plants on tarps in a protected area to keep maturing. Turn these plants one or two times a day to keep them from composting.

Seed extraction

Amaranth panicles are prickly and they get pricklier as they dry down. Put on gloves for your seed work. You can strip the seeds off the dry panicles with your gloved hand or you can hit the stalk against the side of a bin to thresh them.

Seed cleaning

At first glance, amaranth seed can be overwhelming to clean. It is so small and light. But it is actually quite easy to clean.

Start by running your seed and chaff through a quarter-inch screen to remove any larger bits.

Fig. 5.1: *Here's a bin of stripped amaranth panicles.*

Fig. 5.2: *Spread out some of the panicles on a window screen. You can see the seed poking through.*

Fig. 5.3: *Shake the screen and the big bits stay on top. The seeds have disappeared!*

Fig. 5.4: *You wind up with some amaranth seed mixed with a bit of dusty chaff. Time to winnow.*

Then spread the seed mixture over a window screen. The seed will fall through the openings leaving flower bits and small pieces of chaff on top. The seed that comes out the bottom will be clean enough for your uses. A bit of winnowing with a fan on a low setting will clean up the rest.

CALENDULA

The only thing frustrating about growing calendula for seed is that there are two sizes of seed. But it's really only frustrating if you're growing bulk seed. It's easy to get enough seed of any size for your own personal use.

OP availability

There are almost only OP calendula varieties out there.

Managing cross-pollination

Calendula is in the *Calendula officinalis* species and is a crosser. One thousand feet between calendula varieties should be enough to see minimal crossing.

Seed Calendula at Tourne-Sol

We grow calendula as a cut flower for our bouquets. But for many years, we also grew calendula to harvest as flower buds, just as they were opening, to dry and use in herbal teas. In either case, there were enough flowers that would escape or remain post-harvest to produce seed for our own needs.

We also grow bulk calendula seed in our main seed garden. We have cut a few blooms from it when we've been tight in the flower gardens.

Seed yield

Seed Crop Bed Length	Imperial Yield	Metric Yield	Seed Yield
100 bed-feet (30 bed m)	4 to 7 lbs	1.75 to 3.25 kg	84K to 154K seeds
1 bed-foot	0.6 to 1.1 oz	18 to 32 g	800 to 1.5K seeds
1 bed m	2 to 3.6 oz	59 to 105 g	2.8K to 5K seeds

Growing the seed crop

Grow calendula for seed as you would for a cut flower. Leave some of the earlier blooms to make sure you have enough time for the seed to mature.

Seed harvest

Once the petals drop from calendula flowers, you'll see green seeds sitting right there on the flower stem. They're very tempting to harvest but if you go to rub them off the plant, the calendula flower holds tightly onto the seeds. These seeds are not ready.

Wait a few weeks and the seeds turn brown. Now, if you reach for the flowers, the dried brown seeds readily shatter into your hand. Mama calendula is very happy for you to take these seeds and spread them far.

Fig 5.5: *Any calendula blooms you don't pick will readily go to seed.*

Fig 5.6: *The green seeds on the left are not ready. The brown seeds on the right are ready for harvest.*

Fig 5.7: *You can tell these seeds are ready because they shatter so easily when you run your fingers through them.*

For small amounts of seed, you can handpick individual seed heads. For bulk amounts of seed, you can bend the plants over a tarp or wide bin and carefully shake the plants so the dry seed shatters without much green material.

Seed extraction

Calendula seed shatters easily at harvest so there isn't much extra to do for extraction.

Seed cleaning

Calendula seed comes in two sizes: larger pale seeds and darker small seeds. Both sizes are viable but they make it harder to screen the seed effectively, especially with the hooked shape of the seed.

Picking the seed out of the dried flower parts is an easy way to clean small amounts of calendula seed. Even easier is to not clean the seeds at all. Simply put dry seeds and chaff into an envelope. It will be simple enough to sow them this way next year.

You can also use the dry seed techniques from Chapter 8 to clean your seed as you separate the lot into two lots based on size. After cleaning you can mix the two seed lots back together or keep them separate. You choose!

CELOSIA

This was one of the first cut flower varieties that I really fell in love with—as much the cockcomb varieties as the more feathery plumes. For a market garden, you can cut the main flower and then rely on some of the later side shoots for some seeds.

Seed Celosia at Tourne-Sol

At Tourne-Sol we started out by keeping seed from our favorite celosias in the cut flower garden. But these days we have larger celosia plantings where I try to get them to cross up with each other with medium success.

OP availability

Most celosia seed is OP.

Managing cross-pollination

Celosia comes in a number of different flower types but they are all in the *Celosia argentea* species. This means that they can all cross with each other. In my experience, though, celosia behaves more like an intermediate selfer and varieties don't cross readily with other varieties of the same species. Three hundred feet is probably enough space to keep from seeing most crossing between celosias.

Seed yield

Seed Crop Bed Length	Imperial Yield	Metric Yield	Seed Yield
100 bed-feet (30 bed m)	3 to 10 lbs	1.25 to 4.5 kg	1.5M to 4.8M seeds
1 bed-foot	0.5 to 1.6 oz	14 to 45 g	15K to 48K seeds
1 bed m	1.6 to 5.2 oz	46 to 148 g	48K to 156K seeds

Fig 5.8: *As the celosia matures you'll start to see black seeds poking through the flower at the base of the plant.*

Growing the seed crop

Plant your celosia crop as early as possible after your last frost to make sure you have enough time to get to seed. During the vegetative phase of your crop, you can irrigate with sprinklers. But you should avoid any overhead irrigation once the crop starts to go to flower. You can harvest the flowers for bouquets but make sure to leave some branches to go to seed.

Fig 5.9: *We cut full celosia stems and then spread them out in the greenhouse on landscape fabric. We make sure that all the stems are in the same direction.*

Seed harvest

It's time to harvest your celosia once you start seeing black seeds appear on the flowers. You can cut branches or the whole plant above the roots and bring it into a protected dry area to keep maturing for one to two weeks.

Seed extraction

Rub the seed off the plants by hand or hit the plants against the side of a bin to thresh them.

Seed cleaning

Clean your celosia seeds the same way that you would amaranth seeds, using a window screen.

NIGELLA

If you're growing nigella to put the green pods in your bouquets, then you're pretty close to the seed. And if you're growing it for dried pods in dry flower arrangements, then you already have the seed in those pods!

Seed Nigella at Tourne-Sol

My first nigella was Love-in-a-Mist. I discovered this flower growing at Orchard Hill Farm where I was working with Martha and Ken Laing before we started Tourne-Sol. I was captivated by both the wispy flowers and the seed pods. Martha grabbed a couple of pods and told me to just take some seed with me. I crushed the pod and there was so much seed just sitting there in my hand. Now, we grow that Love-in-a-Mist at Tourne-Sol. We use it in our cut flower bouquets as green pods. We can easily get enough seed for our needs from a few escaped blocks, but we also grow full beds when we're growing seed for sale.

OP availability

I've only seen OP nigella out there.

Managing cross-pollination

There are a number of different nigella species used as cut flowers. These are the most common: *Nigella damascena, Nigella hispanica, Nigella orientalis,* and *Nigella sativa.* As far as I've seen, nigella don't cross-pollinate between species. However, if you have two varieties of the same species they can cross. Nigella is a crosser, so if you want to keep varieties true to type, aim for 1,000 feet between varieties of the same species. That being said, multi-colored nigella are beautiful!

Seed yield

Seed Crop Bed Length	Imperial Yield	Metric Yield	Seed Yield
100 bed-feet (30 bed m)	5 to 15 lbs	2.25 to 6.75 kg	1.1M to 3.4M seeds
1 bed-foot	0.8 to 2.4 oz	23 to 68 g	11K to 34K seeds
1 bed m	2.6 to 7.9 oz	75 to 223 g	36K to 111K seeds

Fig. 5.10: *Cut your drying nigella plants and keep them oriented with their heads all in the same way to make it easier to work with them.*

Growing the seed crop

Grow nigella for seed the same way you would for a cut flower. The pods will dry down a few weeks after the flower blooms.

Seed harvest

Cut the stems with the pods and put them in a bin. At this point you could spread them on a tarp to make sure they are nice and dry and ignore them until the fall to deal with. Or you could get the seed out right away.

Seed extraction

For small amounts, crush the pods with your hands. For bulk amounts, put the pods on a tarp and stomp with your boots.

Fig. 5.11: *It's easy to harvest a few nigella seed pods to get a bunch of seed for next year.*

Seed cleaning

Nigella is very easy to clean using the dry seed techniques in Chapter 8.

POPPIES

Poppies are another flower that is just waiting to turn into a seed pod. A few seed pods will give you more poppy seeds than you could ever plant.

Note: What I've written below doesn't necessarily apply to California poppies. These behave quite differently as a seed crop than other types of poppies.

Seed Poppies at Tourne-Sol

We grow poppies to put green pods in our flower bouquets. That means we're mainly looking for big fat pods and the flower color doesn't matter so much. Which is unfortunate because our poppies have a lot of color and frills. In the early years of our farm, I got all the different types of poppies I could and let them cross up to create this mix.

OP availability

Poppy varieties seem to all be OPs.

Managing cross-pollination

Papaver nudicaule, Papaver rhoeas, and *Papaver somniferum* are the three main poppy species you'll find as cut flowers. They do not cross together.

With their big flowers, I assumed poppies would be crossers. But in my experience *Papaver rhoeas* and *Papaver somniferum* act like selfers. When I was trying to get all my poppy selections to cross, I found it frustrating how little poppies would cross even when they were grown side by side. I would only find a couple of crossed plants out of hundreds when I would later grow out their seed. I don't know if *Papaver nudicaule* is as much of a selfer.

Seed yield

Seed Crop Bed Length	Imperial Yield	Metric Yield	Seed Yield
100 bed-feet (30 bed m)	3 to 6 lbs	1.25 to 2.75 kg	8.4M to 16.8M seeds
1 bed-foot	0.5 to 1 oz	14 to 27 g	84K to 168K seeds
1 bed m	1.6 to 3.3 oz	46 to 89 g	269K to 554K seeds

Growing the seed crop

Grow your poppy seed crop the same way you'd grow poppies for cut flowers. Leave some flowers to turn to pods.

Seed harvest

Poppy pods are ready to harvest when they turn white/grey. They will often rattle a bit when you shake them. Cut stems with secateurs and put poppy seeds into bins in the same orientation. Be careful to keep the poppy pod facing upwards during harvest. Some poppy pods have slits along the edge side of the top where seeds can slip out if you tilt them too horizontally.

Fig. 5.12: *If you look closely you'll see one flagged pod in this field of diverse poppies. That was a flower I particularly liked. I'll harvest that pod separately and grow it out on its own next year.*

Seed extraction

Start by turning your poppy pods upside down over a bucket and shaking them. If there are slits along the top edge of the pod, the seeds will shake out. You can also crush the pods in another bin with your hands or boots to see if there are any more seeds to release.

Seed cleaning

Shaken-out seed is pretty clean. Seed from crushed pods will need more cleaning. Start with a window screen just like cleaning amaranth seed. You can winnow poppy seeds with fans but be very careful because the seed is so light.

Fig. 5.13: *When I grow out a crossed-up poppy, I'm hoping to see more of the stuff I like.*

Fig. 5.14: *When the poppy pods have dried down to this point, it is time to harvest the pods for seed!*

Seed Strawflowers at Tourne-Sol

We grow a couple of strawflower varieties with distinct colors for flower bouquets. As I'm typing this up, I realize that we don't actually grow any of these out for seed. Instead we have a crossed-up mix of strawflowers that we offer in our seed catalog. The strawflower seed patch might be my favorite place to walk through on our farm. I'm going to have to start growing some of these distinct strawflower varieties for seed in the years when we don't grow our crossed-up mix.

STRAWFLOWER

This is the last of the really easy flower seeds in this book. If you're growing strawflowers for bouquets, you can easily get enough seed for your needs.

OP availability

Strawflower varieties are all OP.

Managing cross-pollination

Strawflowers are in the *Xerochrysum bracteatum* species. They are crossers, so aim for 1,000 feet between varieties to see next to no crossing; or grow them closer to each other to get even more beautiful crossed-up blooms.

Seed yield

Seed Crop Bed Length	Imperial Yield	Metric Yield	Seed Yield
100 bed-feet (30 bed m)	2 to 4 lbs	1 to 1.75 kg	1M to 2M seeds
1 bed-foot	0.3 to 0.6 oz	9 to 18 g	10K to 20K seeds
1 bed m	1 to 2 oz	30 to 59 g	34K to 67K seeds

Fig. 5.15: *Strawflower seed is ready when the fluffy pappus begins to pour out of the flower.*

Growing the seed crop

Grow your strawflower seed crop the same way you'd grow strawflowers for cut flowers. Leave some flowers to go to mature seed.

Seed harvest

Strawflowers clearly tell you when they are ready—the inside of the flower gets all fluffy and you can gently rub that fluff away to reveal a tiny cup full of seed.

For small amounts of seed, remove the fluffy pappus and pour the seed into an envelope or small container.

For larger amounts, pull off the seed heads by hand with a tugging motion and drop into a bin. Spread them out on a tarp on a wide container in a dry area. At this point, you can leave these to deal with later in the season.

Seed extraction

While the seed heads are lying on a tarp in our seed HQ, I routinely go and stir the seed heads to release the pappus. Some of these will blow away before my next stirring visit. You can then take your seed heads and place them on a screen. Shake the seed heads so that the seed falls through and leaves the empty seed heads above. Remove the empty flowers and then go look at your seed.

Fig. 5.16: *Harvest the individual seed heads and brush the pappus out of the flower.*

Seed cleaning

Strawflower seed is usually pretty clean other than the feathery pappus. Put your seed into a narrow container and shake it so that the seed sinks to the bottom. You can scoop off some of the pappus from the top. Screening and lightly winnowing should do the rest of the job.

Fig. 5.17: *You can pour the seeds out of the flower into your hand. A couple flowers will give a lot of seed.*

Fig. 5.18: *We harvest individual flower heads into a bin and then spread them out to dry in the greenhouse while we wait to process them later.*

SUNFLOWER

This is where I switch to harder seed crops for market gardeners. And who would have thought that sunflowers would be first on that list!

With sunflowers, it's not growing the seed that's the problem. Well, it is a bit of a problem because of all the birds and squirrels who want to beat you to the seed harvest. But otherwise the seed part of sunflowers is easy.

The hard part is that so many commercial cut flower varieties are F1 hybrids. If that is what you're used to it's going to be hard to move to OP varieties if you want to keep your own seed. So if you're into F1 sunflowers, then look for other flowers as your first seed crops.

Seed Sunflowers at Tourne-Sol

We grow a lot of F1 sunflowers in our cut flower garden and they are fantastic for their yield and uniformity. They do just what we want them to. For our seed company, we grow a population of OP sunflowers that have a range of petal colors in all the yellows, oranges, and reds. We grow our seed sunflowers 1,000 feet away from the cut flower garden because I don't want them to cross. We have wound up with some seedless sunflowers from crossing with hybrid sunflowers and that is not good for an OP variety!

OP availability

The market standards are all F1 hybrids, but there are still a lot of OPs out there.

Managing cross-pollination

Sunflowers are in the *Helianthus annuus* species. They are crossers, so aim for 1,000 feet between varieties to see next to no crossing. Make sure to isolate your OPs from F1 hybrids by 1,000 feet.

Seed yield

Seed Crop Bed Length	Imperial Yield	Metric Yield	Seed Yield
100 bed-feet (30 bed m)	5 to 20 lbs	2.25 to 9 kg	36K to 148.5K seeds
1 bed-foot	0.8 to 3.3 oz	23 to 91 g	400 to 1.5K seeds
1 bed m	2.6 to 10.8 oz	75 to 298 g	1.2K to 4.9K seeds

Growing the seed crop

Grow your sunflower seed crop the same way you'd grow sunflowers for cut flowers. Just make sure to not cut all the flowers and leave some to go to seed.

Birds will try to get to your seeds before you do. You can cover sunflower heads with mesh onion bags as the flowers start to show seeds.

Seed harvest

The disk of your sunflower will start to change texture as the seeds start to mature and the outer petals will start to turn yellowish. You can rub off the flower parts of the disk to reveal the seed below. At this point you can cut the flowers with as long as a stem as you can get. Harvest the heads into bins and then bring them to spread out on a tarp in a protected dry area. Leave the heads to keep maturing for one to two weeks. During this time you will still need to keep an eye on birds and other critters who want your seed. You can put hardware cloth or window screen material over the heads to create a physical barrier. This might not be feasible for large amounts of sunflower heads.

Fig. 5.19: *When sunflowers start to hang their heads low and lose their petals, it is time to harvest those heads before the squirrels do.*

Seed extraction

Take two sunflower heads and rub the seeds against each other and the seeds will pop off. You can also break the heads apart with your hands.

Fig. 5.20: *You can also tell that sunflower heads can be harvested when you remove the flowers in the middle of the disk and reveal the seeds underneath. At this stage you can let your harvested sunflower heads sit for a couple of weeks to keep maturing.*

Fig. 5.21: *It's time to get the seed off of this sunflower head.*

Seed cleaning

Use the dry seed techniques in Chapter 8 to clean your sunflower seeds.

Similar to what can happen with beans, there are some insects that lay their eggs in sunflower seeds. Put your dry sunflower seeds into a freezer for one week to eliminate any insect eggs. After you remove the seeds from freezing, let your containers sit for a few hours so that they get back to room temperature before you open them. This will reduce the chance of unwanted condensation.

ZINNIAS

This is the other trickier flower seed in this book. There are three challenges market growers might face with zinnias:

1. If you grow different zinnia varieties, they will all cross with each other making it hard to keep colors and flower shapes distinct.
2. Wet climates make it difficult to get zinnias to go to seed.
3. Zinnias can be tough to clean if you don't have much seed work experience.

So these are not usually great first seed crops.

Seed Zinnias at Tourne-Sol

We grow a lot of zinnias for cut flowers at Tourne-Sol and like to have distinct colors. We also have a fairly wet climate that makes it hard to grow good quality zinnia seed. We have had some successful seed lots but not consistently enough to dedicate much space to zinnia seeds.

We still purchase most of the zinnia seed we use on our farm, but this is one seed crop where I want us to improve our seed production techniques.

OP availability

Most zinnia varieties are OP but there are some F1 out there.

Managing cross-pollination

Zinnia elegans is the main zinnia species on market farms. There are other zinnia species out there and as far I can tell they don't cross easily with *Zinnia elegans*. Zinnias are crossers so keep varieties of the same species separated by 1,000 feet to avoid seeing crossing.

Growing the seed crop

Grow your zinnia seed crop the same way you'd grow zinnia for cut flowers. Choose one of your earliest plantings to be your seed crop so it has time to get to mature seed. Use drip irrigation to keep the drying flowers and developing seed from getting wet. Make sure to keep your team from deadheading the flowers you want to go to seed.

Seed harvest

The flowers dry down as the seed matures. When the seed is ready, you should be able to easily tug and remove the petals from the flower.

Test a few flowers when you want to harvest to see if there is any seed matured amongst the leaves. Once you've identified that the seed is ready, you can pop off seed heads into a bucket and spread them out on a tarp in a dry area.

Fig. 5.22: *Zinnia flower heads that have been drying for a month or two.*

Seed extraction

Rub and shake the dried flower heads to liberate the seed without shattering too much of the flower parts.

Seed cleaning

Use the dry seed techniques in Chapter 8 to clean zinnia seeds.

Fig. 5.23: *Crumble up the flower heads into a bunch of dry flower parts.*

Fig. 5.24: *After winnowing with a fan and screening with a half-inch screen you can get seeds that are clean enough for farm use.*

Zinnia seed can feel particularly overwhelming compared to the other flower seeds I've listed. Take your time with each step as you develop your seed skills. As you screen or winnow, put your discarded material into the same clean bin. This way if you feel you've been too aggressive you can go back to that bin see if there is still seed in your discarded material and possibly clean it again.

GROWING OTHER CUT FLOWERS FOR SEED

As you get used to harvesting the seed from some flowers, you'll see that many other flowers are similar, and you'll also notice that other flowers grow very differently.

Here are some general rules when trying to harvest seeds off of a cut flower crop:

- Choose an early planting so that your seed has enough time to mature.
- Don't deadhead all your flowers—you need to leave some earlier blooms to go to seed.
- As the flowers start to dry, poke around in the flower to see if there are any mature seeds.
- Take your time cleaning your seed, and don't worry if the seed is still kind of dirty after your best efforts. It will still probably germinate, and you'll get better at it!

I've already mentioned this—assume that they are all crossers until proven otherwise. And don't worry about crossing!

Chapter 6

Growing Leafy Greens for Seed

Now we get into the seed crops that are a few steps removed from their vegetable stage. Leafy greens have a particularly strong synergy on market farms. As long as you don't cut the growing point on the plant, you can harvest the leaves from these crops multiple times and still get them to seed.

Most leafy greens are annuals. For these crops, the most important thing is to let earlier plantings in the season go to seed rather than later ones. When you're simply a salad grower, it seems like lettuce and brassica greens just go to flower so quickly. But once you're in it for the seed, you realize that crops bolt much slower than you expect and might not have enough time to get mature seed.

A few leafy greens are biennials. You can get a full season of cutting greens off of these and as long as the plants can survive the winter, you'll get seed in year two.

LETTUCE

Lettuce has a lot going for it as a seed crop. It barely crosses and all lettuce varieties are open pollinated. This means that most lettuce you grow from your own seed should look just like its parent.

It can be very frustrating to grow lettuce in a wet climate because of how susceptible lettuce is to all types of diseases. This is true for a market crop, and even more so for a seed crop that is longer in the ground. Some varieties might get you decent baby leaves, or even a decent head but will melt before you can get the plants to flower and set seed.

Good ventilation and humidity control in your seed crop will go a long way to reduce disease opportunities. But you'll also find that some varieties are a lot more disease resistant than others.

Seed Lettuce at Tourne-Sol

We grow our seed lettuce in an unheated greenhouse to keep them away from wet outdoor weather. We grow seed lettuce varieties side by side with no isolation and we only see the odd crossed-up plant. I love those odd crossed-up plants and harvest their seed separately from the rest of the varieties. The first couple of years I grow these crossed seeds, I grow them outdoors to see how they tolerate ambient conditions and to eliminate any that are particularly susceptible to disease. Then we move them indoors around the third or fourth generation.

OP Availability

All lettuce seeds are open pollinated.

Managing cross-pollination

Lettuce is in the *Lactuca sativa* species. It is a very strong selfer. If you grow it without isolation distances you will barely see any crossing.

Seed yield

Seed Crop Bed Length	Imperial Yield	Metric Yield	Seed Yield
100 bed-feet (30 bed m)	2 to 6 lbs	1 to 2.75 kg	0.7M to 2.2M seeds
1 bed-foot	0.3 to 1 oz	9 to 27 g	7K to 22K seeds
1 bed m	1 to 3.3 oz	30 to 89 g	22K to 73K seeds

Growing the seed crop

Lettuce is a hardy annual. Plant your lettuce crop as early as possible to make sure you have enough time to get to seed. During the vegetable phase of your crop, you can irrigate with sprinklers. Avoid any overhead irrigation once the crop starts to bolt.

If you're growing lettuce as a salad green you may harvest the lettuce leaves repeatedly from your crop. Be careful to not damage the growing tip. If you're

Fig. 6.1: *We grow seed lettuce side by side and see minimal crossing.*

Fig. 6.2: *As the lettuce stems bolt and elongate we strip the lower leaves from the plants to improve ventilation.*

growing lettuce heads, then you cannot harvest from the plant and let it go to seed, since you will be damaging the growing tip of the plant. In this case you'll need to leave some unharvested plants to go to seed.

Transplanted lettuces are at a good spacing for a seed crop. Direct-seeded lettuces might be able to get to seed but there is a higher risk of disease in dense plantings, so you should consider thinning lettuce plants to every six to twelve inches. Alternatively, you can dig up lettuce plants that are finished market harvest and replant them in another area to go to seed. (In a wet climate this might be in a greenhouse or tunnel.)

As the lettuce plant stems bolt and elongate, you can remove lower leaves to improve ventilation and reduce the matter that might bolt. Make sure to do this on a warm sunny day, so that the wounds from the removed leaves heal quickly. This is an especially important step in humid climates.

From the moment lettuce starts to bolt, it takes two to three months to get seed

Fig. 6.3: *Little lettuce flower buds reaching for the sky.*

Seed harvest

Lettuce produces tiny little flowers that are like mini-dandelions. And like mini-dandelions they produce fluffy parachutes. The parachute is called a pappus. Once you see these pappi on your lettuce flowers, the seed is ready to harvest.

Fig. 6.4: *When the flowers turn into a fluffy pappus, the seed is ready for harvest.*

Fig. 6.5: *When it's time to harvest your seed lettuce, put on a mask (it is very dusty) and get in there with your bin.*

Seed extraction

You can shake your lettuce plants into a container to harvest the mature seed and leave any yellow flowers to keep maturing for another harvest. Alternatively, you can cut whole plants and lay them on a tarp in a dry area for the rest of the seed to mature. After one to two weeks, you can shake the dry plants to extract the seed.

Fig. 6.6: *Take the lettuce plant and shake it into a harvest bin. Hit the sides of the bin to help dislodge mature seeds. Be gentle with how you hold the plant; you will come back later for future shaking.*

If you're growing lettuce seed outside in areas with a lot of humidity, you should go for multiple harvests to get seed as soon as possible and not risk your crop succumbing to severe rain events.

You'll wind up with bins of lettuce seed and fluff. Spread this out in a dry area for a few days before cleaning the seed.

Seed cleaning

It can be hard to get lettuce perfectly clean. If you only need a bit of seed, you can sacrifice a lot of your seed harvest to get the rest clean. But with the right screen size and making sure the seed is dry, you can get a lot of the bits out.

Fig. 6.7: *After the seed and fluff has had a few days to dry you can clean the seed with a* $3/64 \times 5/16$ *screen.*

Fig. 6.8: *You should be able to get seed clean enough for your own use this way. There still might be a few bits to get out if you plan on selling this seed.*

Begin by screening with a quarter-inch screen to remove most of the bigger stuff then use a $3/64 \times 5/16$ screen for your last screening to get lettuce clean. This size screen needs to be ordered from a screen supplier. If you start growing a lot of lettuce seed, you will find this screen size is a great investment.

Fig. 6.9: *You can cut your arugula and ...*

BRASSICA GREENS

I've lumped a number of crops together under the moniker brassica greens: arugula, tatsoi, mizuna, mustards, and a number of others. These crops can be divided into three species: *Brassica rapa, Brassica juncea, Eruca sativa.* No matter the species, brassica greens can be grown in a similar way for seed.

I'm always surprised there aren't more salad growers growing all their own salad green seeds. It is easy enough to harvest out a three-to-five-year seed supply from the same bed from which you just cut multiple salad harvests. There is definitely the possibility of cross-pollination, but that is only if you have more than one brassica of the same species in flower at the same time.

Seed Brassica Greens at Tourne-Sol

It's been years since we've bought brassica greens seed for our farm. We only buy a variety once and it goes straight into seed production for us. We start seed brassica greens in the nursery and then plant them out so we have a head start on the season. But if we have a problem in the nursery, we will direct sow them in the field.

We aim for 1,000 feet between varieties, but we've had pretty good isolation with 600 feet between varieties. I love finding occasional crossed-up brassicas. (More on that in Chapter 9.)

Fig. 6.10: *then let it go to flower and seed.*

OP availability

Arugula varieties are pretty much all OP. *Brassica juncea* mustards are mostly OP. There are some F1s out there but the OPs are fantastic. *Brassica rapa* depends on the variety. Tatsoi, Tokyo Bekana, mizuna, komatsuna, and the other brassica greens used as salad greens are mostly OPs. Many bok choys are F1s—trial some OPs to see how they resist bolting and form nice heads at different times of the year.

Managing cross-pollination

Arugula is in the *Eruca sativa* species. Spicy mustards are in the *Brassica juncea* species. The rest are mostly *Brassica rapa*: bok choy, tatsoi, Tokyo Bekana, mizuna, komatsuna, rapini, bok choy, choy sum, and hon tsai tai.

You can grow one of each species without crossing. These are all crossers and if you grow more than one variety of the same species with less than 300 feet distance you will see 20–50 percent crossing. It is very hard to maintain distinct varieties under these conditions. Of course these crossed-up greens can be great for unique salad mixes. A thousand feet between varieties of the same species should avoid crossing.

Do note that only crops in flower will cross. You can grow a number of different *Brassica rapa* for salad mixes, and as long as you till them into the ground before they flower, you'll be able to avoid crossing with a seed crop.

Seed yield

Seed Crop Bed Length	Imperial Yield	Metric Yield	Seed Yield
100 bed-feet (30 bed m)	5 to 10 lbs	2.25 to 4.5 kg	0.9M to 1.8M seeds
1 bed-foot	0.8 to 1.6 oz	23 to 45 g	9K to 18K seeds
1 bed m	2.6 to 5.2 oz	75 to 148 g	29K to 57K seeds

Growing the seed crop

Plant your brassica greens crop as early as possible to make sure you have enough time to get to seed. You may harvest the leaves repeatedly from your crop, just make sure to not damage the growing tip.

During the vegetable phase of your crop, you can irrigate with sprinklers. But you should avoid any overhead irrigation once the crop starts to bolt.

Brassica greens can handle much tighter spacing than lettuces and get a seed crop without much disease problems. If you do grow five to seven rows per bed you might find that is a bit tight. You could pull out some rows to leave two to three rows. If you want, you can also thin plants in row to every four to six inches. If you're concerned about ventilation, plant your brassica greens seed crop a little less densely.

Once your brassicas have bolted, you might consider staking the plants to keep them from contact with the ground (but this isn't a vital step).

Fig. 6.11: *Green brassica siliques at this stage are not quite ready for harvest.*

Seed harvest

Brassica seed pods are called siliques.

Brassica siliques turn brown-golden when the seed matures. For small amounts of seed, you can shake the dry plants into a bin or bag as the plants are still standing. For bulk seed, you should cut the plants above the roots and windrow your whole plants in a dry area.

Some brassica seed shatters very easily, and if you wait too long you'll lose a lot of seed in the field. This has been my experience with arugula. Aim to harvest whole brassica plants earlier in the day when there is still dew to keep the siliques shut.

Fig. 6.12: *By the time they turn a golden brown, it's harvest time. At this point you could simply go through the planting and shake the brassica seeds into a container (similar to lettuce).*

Fig. 6.13: *Or you can cut plants and let them dry on some landscape fabric in a covered space.*

Fig. 6.14: *When everything is nice and dry, have a dance party to stomp those siliques open. Remove the larger plants into a compost pile and gather what's left into bins (this will be seeds, split siliques, and a lot of chaff).*

You can also harvest plants while the siliques are still on the green side. If the seeds are starting to be defined in the pods, you can harvest the plants, then spread them out in the nursery. Let the plants wait a couple of weeks to keep maturing. Turn the plants every day to keep them from composting.

Seed extraction

Spread your plants on a tarp and stomp on them. Brassica siliques don't need much pressure to split in half. After you've stomped on the plants, you can remove the loose material on top (mostly branches and empty siliques). Gather up the seed and chaff left on the tarp into bins and then proceed to seed cleaning.

Seed cleaning

Use the dry seed techniques in Chapter 8.

Fig. 6.15: *Set up a screening station right in the field to remove all the big material from the seed.*

Fig. 6.17: (Above) *Then you can winnow the rest of the seed to get it nice and clean.*

Fig. 6.16: (Left) *This is just a rough screening to reduce the volume of material you're handling. Use a quarter-inch screen to remove siliques and chaff from seed. You'll wind up with a bin of seed and smaller chaff.* Credit: Emily Board

DILL AND CILANTRO

If you grow dill or cilantro for bunched herbs then you know how quickly they'll go to seed. You can get a couple of cuts and then you can harvest enough seed for years.

What I find a bit challenging with growing my own dill and cilantro is that the germination rates are not always as high as I would like them to be. If you have this problem and you're growing seed for your own farm, then you can increase your seeding density to make sure you get the stand you're looking for.

Dill and cilantro seed is also pretty cheap to buy so I hesitate to recommend this as your first planned seed crop. Still, this is a great spontaneous seed crop that you can harvest.

Seed Dill and Cilantro at Tourne-Sol

We sow dill and cilantro every two weeks to have a steady supply of fresh herbs throughout the season. We let many of these seedlings go to flower to create a nectar source for all the pollinators in the neighborhood. This also gives us dill flowers to use when making pickles or for flower bouquets. We till most of these plantings under just before the flowers start to turn to seed, occasionally leaving a planting to go all the way to seed.

OP availability

There are only OP varieties for dill and cilantro.

Managing cross-pollination

Cilantro is in the *Coriandrum sativum* species and dill is in the *Anethum graveolens* species. They are both crossers but they don't cross with each other. Most market gardeners only grow one variety of each so crossing isn't really an issue.

Seed yield

Seed Crop Bed Length	Imperial Yield	Metric Yield	Cilantro Seed Yield	Dill Seed Yield
100 bed-feet (30 bed m)	10 to 20 lbs	4.5 to 9 kg	336K to 693K seeds	1.8M to 3.7M seeds
1 bed-foot	1.6 to 3.3 oz	45 to 91 g	3.4K to 6.9K seeds	18K to 37K seeds
1 bed m	5.2 to 10.8 oz	148 to 298 g	11K to 23K seeds	58K to 121K seeds

Growing the seed crop

Plant your cilantro/dill crop as early as possible to make sure you have enough time to get to seed. During the vegetable phase of your crop, you can irrigate with sprinklers. But you should avoid any overhead irrigation once the crop starts to bolt. You may harvest the leaves repeatedly from your crop.

Fig. 6.18: *The middle bed in this greenhouse is planted to five rows of cilantro in the first half of the bed and dill in the second half. This is an early planting intended for fresh bunching.*

Cilantro and dill can handle quite tight spacing and get a seed crop without much disease problems. The plants can get a bit unruly and you can trellis them if you want them to stay under control.

Seed harvest

Cilantro and dill seeds are mature when the seeds on the umbels turn from green to brown. You can harvest these seeds by shaking them off of the plant into a bin. You can also harvest individual umbels or whole plants and bring them in to keep drying on a piece of landscape fabric.

Fig. 6.19: (Above) *The cilantro goes to flower.*

Fig. 6.20: (Top right) *As does the dill.*

Fig. 6.21: (Right) *The dill plants are a bit unruly when they go to seed. They could probably benefit from a little trellising.*

Seed extraction

Thresh the plants by hitting them with a stick. They thresh quite readily.

Try to rub the umbels as little as possible to reduce the amount of small sticks that will shatter and mix in with your seeds. This will make seed cleaning a lot easier.

Seed cleaning

Use the dry seed techniques in Chapter 8.

SPINACH

If you grow spinach as a market green, you might be able to get a good seed crop, but not necessarily so. This is because spinach flowers at specific day lengths, prefers cooler weather for good pollination, and doesn't want to be too wet or too dry for the seed crop to thrive.

Spinach is probably not a great first seed crop, but it definitely could be a good year three or four seed crop.

OP availability

There are increasing amounts of F1 market standards out there but there are still some good OPs available.

Managing cross-pollination

Spinach is *Spinacia oleraceae* and is a crosser and will cross up readily. If you grow spinach varieties that have similar leaf types, you might not care much about crossing. But if you grow both savoyed spinach and smooth spinach, you probably don't want them crossing. Keep them at least 1,000 feet apart to see little crossing. This distance is assuming you or folks in your neighborhood are not growing huge plantings of spinach for seed. Spinach is wind pollinated, so if you're in an area where spinach seed is grown on a large scale, you might not be able to grow spinach seed without it getting crossed up.

Fig. 6.22: *These beds produced half a bin each of dill and cilantro seed.*

Seed Spinach at Tourne-Sol

Spinach has been a challenging crop for us both as a market crop and as a seed crop. Our climate is a little too hot and a little too wet for what spinach would prefer. Our best spinach seed crops have been sown in the fall and overwintered in tunnels or in the field. These plants seem to get much better established before they go to seed. We have been able to cut these plants for vegetable harvest in the fall and sometimes even in the spring before they bolt.

Seed yield

Seed Crop Bed Length	Imperial Yield	Metric Yield	Seed Yield
100 bed-feet (30 bed m)	6 to 14 lbs	2.75 to 6.25 kg	230K to 529K seeds
1 bed-foot	1 to 2.3 oz	27 to 64 g	2.3K to 5.3K seeds
1 bed m	3.3 to 7.5 oz	89 to 210 g	8K to 17K seeds

Growing the seed crop

You can grow spinach as an annual or as a biennial. The important thing is that spinach needs 13–15 hours of sunlight to bolt. It also needs cooler temperatures to pollinate so that viable seed can be produced. This can cause a problem in warmer climates where it is already hot when the days get longer.

To grow spinach as an annual, plant your spinach crop as early as possible to make sure you have enough time to get to seed. You may harvest the leaves repeatedly from your crop. Just make sure to not damage the growing tip.

To grow spinach as a biennial, plant the crop in the fall and let it overwinter. In the coldest climates, you might cover spinach with a mini-tunnel or some row cover to ensure winter survival. But I've been impressed how hardy spinach can be with only snow cover during cold winters.

You can irrigate your spinach with sprinklers while it is a leafy green but you should avoid any overhead irrigation once the crop starts to bolt. Ideally spinach seed plants should be six to twelve inches apart in the row. However, spinach can handle much tighter spacing than lettuces and get a seed crop without many disease problems.

Fig. 6.23: *As your spinach plants bolt, the pollen-producing plant will get taller first. The shorter plants will be the seed-producing plants.*

In my experience, fall-planted spinach produces bigger plants before they bolt and usually yields more seed than spring-planted spinach.

Be aware that there are two types of spinach plants:

- Some plants produce pollen and no seed. These have less leaves and more obvious flowers. If you shake these plants when they are in flower, you will see pollen be released.

- Other plants receive pollen and produce seeds. These plants have many leaves as they bolt. You can see little balls in the crooks of the leaves that will turn into the seeds.

You need both types of plants to flower at the same time otherwise you will not get any pollination and no seeds. Once you see that the seeds have set, you can remove the pollen-producing plants.

Fig. 6.24: *Make sure to keep both types of plants as your plants flower.*

Seed harvest

Spinach seed starts green and generally turns a golden-brown color when the seed is mature. There are some frustrating spinach seeds that do stay quite greenish at maturity. If your spinach seed starts shattering, it is definitely time to start harvesting.

Pull spinach plants and bring them into a dry area and lay them on tarps to keep maturing for one to two weeks.

Seed extraction

Use a glove to rub the seeds off the stems. You can also thresh and winnow the seed, but you might find that rubbing the seed off is pretty quick.

Seed Cleaning

Use the dry seed techniques from Chapter 8.

Fig. 6.25: **(Above)** *Once you can see that there is a good seed set on the seed-bearing plants, you can remove the pollen-producing plants.*

Fig. 6.26: **(Left)** *Only seed-bearing plants are left in this spinach population.*

COLLARDS AND KALE

It is now time to tackle the first biennial seed crops in this book. Collards and kale are both biennials. That means they produce leafy greens that you can pick through their first year and then you need to overwinter the plants so they can produce seed in year two.

Biennials can make it more complicated to grow a seed crop, but there are two scenarios where collard and kale are fantastic first seed crops. Scenario 1: when you live in a climate where these plants can survive the winter outdoors with no protection. Scenario 2: when you live in a harsher winter climate but grow fall collards or kale in a tunnel or greenhouse. In both these scenarios, your kale plants will happily bolt and flower in the spring then produce seed.

As the plants bolt, you suddenly get another bonus crop: collard and kale raab. These raabs are similar to rapini but instead of a bitter flavor these flower buds are sweet from all the exposure to cold. You can harvest the main stalk of each plant and a few side shoots and still have tons of flowers to go to seed.

The only challenge will be whether you need to dig plants up in the spring to get them out of the way if you want to use the space for a different crop.

Seed Collards and Kale at Tourne-Sol

We mainly grow kale at Tourne-Sol but collard seed production would look exactly the same.

We plant kale in unheated greenhouses in August to harvest for our fall vegetable baskets. After veggie basket season is over the farm team keeps eating the kale through the winter. When we want a seed crop from these plants, we leave the kale in place until spring and let it go to seed right where it is. These plants lock up growing space until August but it's easier than moving the planting. We trellis the kale crop to keep it out of any neighboring greenhouse crops.

Each seed crop produces enough seed for many years. So we don't need to let kale go to seed every year.

OP availability

There are hybrid varieties of both collards and kale but there are also a lot of OPs out there. You should try some OPs to see if they can fit well into your farming system.

Managing cross-pollination

Collards are in the *Brassica oleracea* species. Kale can be one of two species: *Brassica oleracea* (this includes curly kales and lacinato kales) or *Brassica napus* (this includes Siberian and Russian kales.)

Collards and oleracea kale will cross with each other and with broccoli, cabbage, cauliflower, kohlrabi, and Brussels sprouts. Napus kale will cross with rutabaga. *Brassica oleracea* and *Brassica napus* will not cross with each other so you can grow a napus kale and an oleracea collard without fear of crossing.

Both *Brassica oleracea* and *Brassica napus* are crossers and need to be isolated from varieties of the same species by 1,000 feet to see minimal crossing.

Remember that all these crops need to be in flower to cross. So you can have a garden full of oleraceas and napus, and as long as only one is in flower, there will be no crossing.

Seed yield

Seed Crop Bed Length	Imperial Yield	Metric Yield	Seed Yield
100 bed-feet (30 bed m)	6 to 14 lbs	2.75 to 6.25 kg	230K to 529K seeds
1 bed-foot	1 to 2.3 oz	27 to 64 g	2.3K to 5.3K seeds
1 bed m	3.3 to 7.5 oz	89 to 210 g	8K to 17K seeds

Fig. 6.27: *Kale raab is a delicious spring vegetable. You can harvest the main stalk of each plant and a few side shoots and still have tons of flowers to go to seed.*

Growing the seed crop

Grow your collards or kale crop the same as if they were a market garden crop. You can pick their leaves many times but make sure not to damage the growing point. You can leave flowering plants at the same spacing to produce seed or can remove some plants to improve ventilation for disease control.

Collards and kale are biennial crops that require 10–12 weeks at under 50°F (10°C) to initiate flowering. In cold climates, you can overwinter your crops in the field or in a greenhouse and get enough cold. In warmer climates, you might not be able to get enough cold outdoors. In this case, you will need to dig up plants and bring them into a cold room for 10–12 weeks.

Plants that have overwintered in the field can also be dug up in the spring and moved to a seed garden. This might be especially tempting if plants overwinter in a tunnel or greenhouse and you want to use that space for something else in the spring.

You can stake and trellis bolted plants to keep them from flopping over and getting in the way.

Fig. 6.28: *This is what a greenhouse kale planting looks like in year two. We trellis the plants to keep them out of their neighbors' space.*

Seed harvest, seed extraction, seed cleaning

Pretty much the same as brassica greens. See pages 95 and 96.

SWISS CHARD

There are two reasons to grow chard for seed beyond the seed itself. One is that the bolting plants before they set seed are some of the most beautiful plants with their colored stems and elegant stature. The other is the fragrance of the chard pollen. *Chard pollen, in its fascinating ambivalence, is the aroma of paradox, of yang and yin commingled, of life and death combined in vegetable absolute.* (Actually this is what Tom Robbins wrote about beet pollen, but it is equally true for chard.)

Otherwise chard is another fairly easy seed crop. It doesn't overwinter quite as easily as collards or kale do. And the plants take a lot more space so you'll have to thin out some plants if you're just letting a market planting go to seed. It can also be a bit tough to harvest at the right time to get the best germination, but you can get a lot of seed off a few plants.

As I mentioned in Chapter 1, chard and beets are the same species. And in the beet section in Chapter 7 I make a price-based case why you might want to grow your own beet seed. A similar argument can be made for chard but generally you probably don't use as much chard seed as beet seed so there are likely more savings to be made growing your own beet seed then chard seed.

OP availability

Though there are a few F1s out there, OP chards are fantastic.

Managing cross-pollination

Swiss chard is in the *Beta vulgaris* species just like beets. They are crossers that will readily cross-pollinate with each other. If you grow chard varieties near each other without isolation, you will get a rainbow of colors. For smaller chard plantings, 1,000 feet should keep you from seeing much crossing. This distance is much too small if you are in areas where beet or chard seed is produced commercially. *Beta vulgaris* pollen is quite light and travels far in the wind. In these areas, you will need two to three miles to avoid crossing.

Seed yield

Seed Crop Bed Length	Imperial Yield	Metric Yield	Seed Yield
100 bed-feet (30 bed m)	15 to 25 lbs	6.75 to 11.25 kg	336K to 574K seeds
1 bed-foot	2.4 to 4.1 oz	68 to 114 g	3.4K to 5.7K seeds
1 bed m	7.9 to 13.4 oz	223 to 374 g	11K to 19K seeds

Growing the seed crop

Chard can be started and initially grown like a market crop, but it will ultimately need more space. Mature plants get pretty big and unwieldy, so thinning them or replanting them to one row spaced at 12–18 inches will give you enough room to handle this crop.

Chard is a biennial that requires ten weeks at 40°F (4°C). It will need to overwinter before going to seed. In cold climates this might be possible in the field or they might need to be in a greenhouse or tunnel. In warmer climates, if you don't have enough cold time, you will need to dig up plants and bring them into a cold room to get enough vernalization.

Plants that have overwintered in the field can also be dug up in the spring and moved to a seed garden. This might be especially tempting if plants overwinter in a tunnel or greenhouse and you want to use that space for something else in the spring.

You can stake and trellis bolted plants to keep them from flopping over and getting in the way.

Seed Chard at Tourne-Sol

We grow chard in our greenhouses to harvest for fall veggie baskets. If we leave these plants where they are they will survive the winter. So there is definitely the potential for us to choose the best of these plants and harvest the seed. But we've struggled with getting consistent germination with chard seed, so this has not yet become a main seed crop for us.

Seed harvest

Chard seed is visible on the stems of the plant. They start off a green color and turn to tan/brown as they mature. When 75 percent of the seed on the plant has changed color, pull the plants out of the ground. Cut some of the root off to leave the dirt in the field. Bring the plants into a dry area to continue to mature.

Fig. 6.29: *Majestic chard plants and fragrant pollen.*

Seed extraction

Even at a market garden scale, you can strip the seeds off the plants by hand. Make sure to wear gloves.

Seed cleaning

Use dry seed techniques from Chapter 8 to get your chard seed clean.

ESCAROLE AND ENDIVES

I was confused by these crops for a while. When I think of endives I think of those chicons grown in dark conditions with a delicious hint of bitter. But those are actually called Belgian endives and they are not an endive at all—they are a chicory. What?

Endives and escaroles are leafy greens that don't form much of a tight head. Endives are also called frisées and are used in many salad mixes. Escaroles tend be grown as full heads, usually in the fall when the cold weather mellows their bitterness.

If you grow endives or escaroles you can probably grow them for seed. They are pretty resilient plants and they are selfers, but they can take a long time to go to seed and this can be tough in some climates. They can also be very irritating to clean.

Seed Endives and Escarole at Tourne-Sol

Endives and escaroles are on my "frustrating" list. When I overwinter them outside under mini tunnels they tend to die. When I grow them as annuals they just don't mature seed in time. I've had some success overwintering them in tunnels and unheated winter greenhouses. And then there's the challenge of cleaning the seed. All that to say that I'm still dialing this one in, which is why it's one of the last crops in Chapter 6.

OP availability

Most escaroles and frisée varieties are OP.

Managing cross-pollination

Escaroles and frisée are in the *Cichorium endivia* species. They are selfers. You can grow them quite close without seeing much crossing.

Seed yield

Seed Crop Bed Length	Imperial Yield	Metric Yield	Seed Yield
100 bed-feet (30 bed m)	2 to 6 lbs	1 to 2.75 kg	0.4M to 1.4M seeds
1 bed-foot	0.3 to 1 oz	9 to 27 g	4K to 14K seeds
1 bed m	1 to 3.3 oz	30 to 89 g	14K to 46K seeds

Growing the seed crop

Endives and escarole can be grown the same for seed as they would be for a market crop. During the vegetable phase of your crop, you can irrigate with sprinklers but avoid any overhead irrigation once the crop starts to bolt. You may harvest the leaves repeatedly from your crop. Just make sure not to damage the growing tip.

If you've planted densely for salad greens, you should thin out the planting to make room for these big plants. Ideally endive and escarole seed plants should be two or three rows per bed and spaced 12–18 inches apart in the row, depending on the size of the plant.

Escarole and endives take a long time to go to seed. Depending on your climate you might be able to plant them early in the spring and get them to seed before your inclement fall weather. But you might not.

You can also grow endives and escarole in the fall and overwinter the plants to get seed the next year. Though these are hardy greens, they are not as hardy as spinach or chicories and might need additional protection in your climate.

Seed harvest

It is not always obvious whether seed is mature in escarole plants. The seed hides in little capsules that aren't obvious to the untrained eye. If flowers have dried back for a bit there is probably some seed. Go and crumble the capsules where the petals have long dried up and see if you find seed in your hand. You can also look on the ground to see if there is shattered seed.

Escarole has seed that shatters easily and holds on tightly to the plant. If you only want a few thousand seeds for your own use then you should shake the plants a bit and take what shatters easily and ignore the rest. If you're looking for a bigger seed harvest you can pull the plants and let them dry in a greenhouse for one to two weeks.

Fig, 6.30: *Escarole going to flower.*

Seed extraction

When the plants are dry, you can thresh them. You'll need to hit them with quite a bit of force to get a lot of seed to shatter. It can be frustrating to get a lot of seed from these, so stop while you're ahead.

Seed cleaning

Use the dry seed techniques in Chapter 8.

Chicory seed is on the small side and at first you might not be able to tell the seed from the chaff, but some winnowing goes a long way.

RADICCHIO AND OTHER CHICORIES

Chioggia radicchio (those red round radicchios) is one of the more common types of chicory in North America, but it is just the tip of chicory diversity. Castelfranco radicchio, sugarloaf, puntarelle, and tardivo di Treviso are just some of the very different chicories out there. (Not to mention Belgian endive which is not an endive.)

These are my favorite vegetables (I look forward to every autumn and that first bitter refreshing bite into a sugarloaf or radicchio head), but they are also a bit more difficult to grow for seed than the other leafy greens in this chapter. Overwintering and seed harvest/extraction can both be challenging. However, this is one crop that will really respond to your seed work. Selecting plants that thrive well in your conditions means better chicories for your farm.

This is only a first seed crop if you are obsessed with chicory and truly know that #bitterisbetter.

Seed Chicory at Tourne-Sol

In 2011, we started growing all the chicories we could find. We planted them in midsummer and wanted to see if they could produce nice heads by fall. There was so much diversity in all these varieties and many never had time to head up before winter came. It was pretty clear to us which varieties had more potential on our farm. But it has still taken us a long time to get reliable chicory seed since then.

Overwintering has always been the biggest problem for us. We started by digging up our favorite plants and planting them into pots in our cold room. They'd look great when we brought them in but by the end of the winter most of the plants in the cold room would be dead. This left us far too few plants to get much seed and we'd try again the next year.

After many years of not being happy with replanting chicory, we started overwintering them in mini-tunnels in the field. This was less work than digging up plants and was a bit more successful. But we had a lot of winterkill during more extreme winters.

In recent years, I've noticed something that I should have seen sooner. Even when a chicory dies back completely above ground, it will often regrow from the roots in the spring if we're patient enough and don't till the ground where the plants were. Sometimes we only see what chicories have survived at the end of May.

I'm only starting to harness these new realizations to be able to grow bulk chicory seed, but I'm pretty excited for out next harvest!

OP availability

There are a lot of OPs and F1s out there. Some OPs can be quite variable. Try a number of varieties to see what does well in your climate. A bit of selection can quickly improve a chicory variety.

Managing cross-pollination

Chicory is in the *Cichorium intybus* species They are crossers and will cross-pollinate readily if grown closer together. Chicories are the only greens that I prefer to not have cross up since you can lose that nice tight head they produce. You should grow chicories with 1,000 feet isolation to see very little crossing.

As mentioned above, Belgian endive is also in this species, as are the wild chicory weeds that grow in many areas. So you should be aware of any of these flowering close to your seed radicchios. And to make this a bit more complicated, chicory can be cross-pollinated from the pollen of endives and escaroles, which are in the *Cichorium endivia* species. (Endives and escaroles are a lot less likely to get cross-pollinated by *Cichorium intybus*.)

Seed yield

Seed Crop Bed Length	Imperial Yield	Metric Yield	Seed Yield
100 bed-feet (30 bed m)	2 to 6 lbs	1 to 2.75 kg	0.4M to 1.4M seeds
1 bed-foot	0.3 to 1 oz	9 to 27 g	4K to 14K seeds
1 bed m	1 to 3.3 oz	30 to 89 g	14K to 46K seeds

Growing the seed crop

You can grow chicory for seed the same as you would for a market crop. If you're growing chicory for salad greens, you may harvest the leaves repeatedly from your crop—just make sure not to damage the growing tip. If you're growing for full heads, you won't be able to cut full heads and still keep the growing tip intact. During the vegetable phase of your crop, you can irrigate with sprinklers. But you should avoid any overhead irrigation once the crop starts to bolt.

Chicory is a biennial. Plant chicory in mid- to late summer depending on where you are. You want enough time so that you are able to see your chicory mature before the crop stops growing due to extreme cold temperatures. In milder

Fig. 6.31: *In the fall I make sure to look at every plant to see if it has a nice tight head or not. I remove all the plants I really don't like.*

Fig. 6.32: *I flag my favorite plants to be my stock seed. The rest of the plants will be our bulk seed. Then we set up arches.*

climates, chicory will probably survive the winter without protection. In colder climates, you might need to protect your chicory with tunnels though it's possible tunnels aren't enough protection in very cold climates for your chicory heads to make it through the winter. However, even with total above-ground winterkill, I have been surprised to find chicory plants start to sprout back from the roots.

In year two, chicory turns to sprawling plants with many branches. You might want to trellis these to keep them orderly.

Seed harvest, seed extraction, seed cleaning

Follow the same instructions as for endives and escarole. See page 107.

Fig. 6.33: *Cover this all in a tunnel if you have very cold winters.*

Fig. 6.34: *In a best-case scenario, chicories look like this when they get out of winter.*

Fig. 6.35: *I peel back some of the gunky leaves to reveal radiant leaves.*

Fig. 6.36: *In the worst case the plant completely dies and then in mid-May I see a little sprout like this pop up.*

Fig. 6.37: *Chicory can be very sprawly. Trellising can keep things a little more orderly.*

Chapter 7

Growing Roots and Other Such Vegetables for Seed

AND SO WE GET TO ROOT VEGETABLES. I use the term "root" to include a number of crops that are not technically roots but are still treated this way. There are bulbs such as onions. And swollen hypocotyls (the hypocotyl is the bottom part of the stem) such as turnips, rutabagas, or radishes.

At first glance root vegetables are more complicated than the other crops in this book because you need to figure out how to get them to overwinter. But once you've developed an effective wintering strategy, getting root vegetables to seed in year two isn't much different from getting many leafy greens to seed.

Root vegetables offer some unique opportunities on market farms that you might not find if you were only growing for seed. Market growers often handle thousands if not tens and hundreds of thousands of plants of a crop. This gives you a chance to see many different individuals and be able to select the very best to go to seed. Then you can sell all the rest. Choosing the best hundred individual roots out of ten thousand roots will probably get your variety moving in the direction you want and can have a big impact over a few generations.

SPRING AND WINTER RADISHES

Spring radishes are your small round and French breakfast radishes that are ready for bunching in three to four weeks from sowing. You can plant them all summer if you treat them right. They have become a staple of small market gardens and you can use a lot of seed if you have regular successions through the growing season.

Winter radishes are the larger radishes and daikon. They are planted in midsummer for fall harvest. They store very well through the winter. You probably grow less winter radishes than spring radishes, and so use less winter radish seed (especially since they are seeded less densely).

Spring and winter radishes are very similar with regard to being the easiest root crops to get to seed, but it can be frustrating to clean the seed if you don't have the right techniques.

Seed Radishes at Tourne-Sol

We sow spring radishes for seed in April in one of our unheated greenhouses. These are seeded as densely as if we were growing them for market. We pull all the radishes and look at them one by one. We select the best radishes for seed and we eat any radishes that don't make the cut, enjoying the first taste of spring. We put the selected seed radishes in the cold room for two weeks and then plant them out in the field.

We sow winter radishes in the field around August first and harvest them in mid- to late October. At harvest, we pull all the radishes and select the best 20–50 percent for seed and the rest go into our veggie baskets or into the kimchi pot.

OP availability

There are many F1 spring radishes out there but there are also great OP spring radishes as well. Most winter radishes are OP.

Managing cross-pollination

Spring radishes and winter radishes are both *Raphanus sativus* and can cross together. Usually winter radishes flower earlier in the season than spring radishes which means you can grow both with minimal isolation.

All radishes are crossers. Grow them 1,000 feet from other radishes to see little crossing. Be careful that any of your market radishes don't go to flower and cross with your seed radish crop. If you have wild radish in your area, they can also cross with your radishes.

Seed yield

Seed Crop Bed Length	Imperial Yield	Metric Yield	Spring Radish Seed Yield	Winter Radish Seed Yield
100 bed-feet (30 bed m)	5 to 10 lbs	2.25 to 4.5 kg	240K to 480K seeds	160K to 320K seeds
1 bed-foot	0.8 to 1.6 oz	23 to 45 g	2.4K to 4.8K seeds	1.6K to 3.2K seeds
1 bed m	2.6 to 5.2 oz	75 to 148 g	8K to 16K seeds	5K to 10K seeds

Growing the seed crop

Grow your seed radishes the same way you would for a market crop. At harvest, pull your radishes and place them on the ground so you can inspect them. Choose your favorite radishes for seed and you can sell the rest. Be careful to leave one inch of greens on your seed radishes so you don't damage the growing point.

There are a few differences between growing spring radishes and winter radishes.

Spring radishes are annuals. Sow them early in the season since they can take a frustratingly long time to reach mature seed in shorter seasons. These can be replanted immediately at six-inch spacing in two rows per bed, and let go to flower. However, spring radishes don't always bolt at the same time. One way to synchronize flowering is to put your harvested spring radishes in the fridge/cold room for one to two weeks.

Winter radishes are biennials. They need temperatures under 40°F (4°C) for 12 weeks. It is easy to provide enough cold if you store your radishes in a cold room at 32–40°F (0–4°C) through the winter. In the spring, plant radishes into three-inch pots in your nursery to let them break dormancy. After two weeks in pots, your radishes should have put on some nice foliage and look vibrant. Discard any roots that don't look good. Transplant the winter radish from the pots into the field with two rows per bed and one-foot in-row spacing.

In some climates you can simply let your radishes overwinter outside in the ground. In this case you should still dig up the roots to inspect them and then replant them.

Once spring and winter radishes have been replanted, they can be treated similarly. They both have surprisingly large bushy plants for such small vegetables. You can

Fig. 7.1: *Pull all your radishes and choose the best for seed.*

Fig. 7.2: *Replant your radishes with 6–12 inches in-row spacing.*

Fig. 7.3: *Plant your winter radishes from the cold room into three-inch pots in the nursery.*

Fig. 7.4: *Only transplant to the field radishes that have good roots and good leaves.*

stake the bolting plants to keep them standing up. Radishes will take two to three months to get to seed once they are replanted.

Seed harvest

Radish seed pods are called siliques. The seed is ready when the silique turns from green to a white or tannish color. You can break some pods to see if there is mature seed inside. Radish seed does not shatter easily and is somewhat resistant to precipitation. If your plants are staked, you can leave them in the field until almost all the siliques are mature.

When it's time to harvest, cut the plants just above the top of the root and place the whole plants on landscape fabric. If your plants aren't fully dried, you can clean them right away if you wish. Or you can put the plants in a dry area to deal with later.

Seed extraction

Getting radish seed out of the siliques can make for a miserable time. The pods are foamy and squishy and don't readily shatter like other brassica siliques do.

Here are two secrets to make radish seed extraction easier:

First, make sure that your radish plants are extra dry when you go to thresh them. Also, you should only thresh radish on dry days. Any ambient humidity can make your radish seeds resistant to threshing.

Second, shove your radish plants in a bin and then crush whatever you can with a series of kicks and stomps. Screen out the seed and smaller bits to deal with separately. Pour the remaining intact siliques off the top of the screen back in the bin for another stomping round. Repeat until you've threshed everything.

If you have a seed thresher, it can really make extracting radish seed a lot easier.

Seed cleaning

Once radish seed is extracted it is easy to use dry seed techniques from Chapter 8 to get the seed clean.

Fig. 7.5: *Trellis your radishes to keep them nice and tidy as they flower.*

Fig. 7.6: *You can let your radish pods dry down in the field. They won't shatter like other brassicas do.*

Fig. 7.7: *Put dried plants into a bin and stomp to shatter the siliques. Screen out the shattered seed and then stomp the unshattered siliques again.*

Fig. 7.8: *Screen and winnow the seed from the shattered siliques.*

TURNIPS AND RUTABAGA

There are two very different vegetables out there that get called turnips. On one hand is a big, dense, yellow-fleshed vegetable that many folks would never consider eating raw. These are also called rutabagas. On the other hand is a smaller white-fleshed vegetable that is delicious raw. The outside skin comes in all different colors—completely red, completely white, white with a purple top.

I don't care what you call these crops but just know that rutabagas and turnips are two different species so you need to be clear on what species you're dealing with when you're growing them for seed.

Turnips are a great seed crop for market growers because they are easy to grow, store, and get to seed and you probably use a lot of turnip seed on your farm. Rutabaga are an easy crop to get to seed but most growers only use a small amount of seed any year so they are probably not a great first seed crop.

Seed Turnips and Rutabagas at Tourne-Sol

We sow turnips in late July to mid-August for a fall harvest. We'll select the best plants for seed and send the rest to our vegetable baskets. We overwinter turnips in our cold room and plant them out in the field. Sometimes we plant turnips into three-inch pots in the nursery to break dormancy, but turnips seem so excited to start growing, we usually just plant them straight in the field without problem.

To be honest, we no longer grow rutabaga as a market crop. When we did, we would start the seedlings in the nursery and plant them in the field in mid-June to early July, aiming for a fall harvest. We would choose the best rutabaga for seed. One rutabaga seed crop could produce enough seed for us for many years of farm use and sale.

OP availability

There are some great OP turnips and rutabagas that should meet your seed needs. The one exception that everybody loves are Hakurei turnips which are F1 hybrids.

Managing cross-pollination

Turnips are *Brassica rapa* and will cross with other *Brassica rapa* varieties (see brassica greens on page 94 for a lot more about *Brassica rapa*). Rutabaga are *Brassica napus* and will cross with other *Brassica napus* (mainly Siberian and Russian kale). Turnips and rutabaga will not cross with each other.

Turnips and rutabagas are crossers. Grow them 1,000 feet from other crops of the same species to see little crossing.

If you are in canola country, be aware that canola can be either a *Brassica rapa* or a *brassica napus*.

Seed yield

Seed Crop Bed Length	Imperial Yield	Metric Yield	Turnip Seed Yield	Rutabaga Seed Yield
100 bed-feet (30 bed m)	5 to 10 lbs	2.25 to 4.5 kg	0.9M to 1.8M seeds	0.7M to 2.2M seeds
1 bed-foot	0.8 to 1.6 oz	23 to 45 g	9K to 18K seeds	7K to 22K seeds
1 bed m	2.6 to 5.2 oz	75 to 148 g	29K to 57K seeds	23K to 71K seeds

Growing the seed crop

Turnips and rutabaga are both biennials. Grow the roots as you would for a normal market crop but time your plantings so that you don't have gargantuan turnips or rutabagas in the fall.

Select the best roots for seed and send the rest to market. Make sure to leave about an inch of green at the top to avoid cutting into the growing point. Do not remove roots on the bottom—they won't grow back if you do.

Store turnips and rutabaga in your cold room at 32–40°F (0–4°C) for the winter and plant them out when you can work the ground in the spring. In the second year, plant turnips back in the field with two to three rows per bed and one foot between plants. Plant rutabaga back in the field with one row per bed and one foot between plants.

In some climates you can simply let your turnips or rutabaga overwinter outside in the ground. In this case you should still dig up the roots to inspect them and then replant them.

The plants for both these crops are tall and often pretty sturdy. You can stake and trellis the plants if they are in the way or if you want to keep them nice and tidy.

Turnips and rutabaga will take two to three months to get to seed once they are replanted.

Seed harvest, extraction, and cleaning

All this seed work is the same as for brassica greens on page 95.

ONIONS

Growing onion seed is as EASY as popping those sprouting onions into the ground and letting them put on green leaves and then flowers, which in turn become seeds. And growing onion seed is as HARD as the amount of rain you get.

Heavy precipitation and hail and harsh winds can spread disease through your onion fields over the course of a weekend, creating perfect conditions to compromise your seed crop. If you want to grow onion seed in a wet climate, you should grow your onion seed in a tunnel or greenhouse.

Onion seeds have the reputation of only being good for one year. In my experience this can be true for onion seed bought from seed companies. But farm-grown onion seed is often good for longer than that. This can be a reason to grow your own seed.

This seed crop can be a good first or second crop if you can provide dry conditions for your flowering plants.

Seed Onions at Tourne-Sol

At Tourne-Sol, we choose the best onions at harvest and store the bulbs in our cold room through the winter. We plant the onion bulbs into an unheated greenhouse in early April and let them go to flower. Seed harvest is about mid-August.

OP availability

There is an increasing number of F1 onions on the marketplace, especially for storage onions. But there are still some solid OPs available. Make sure to trial different OP varieties to see how well they perform and how well they store.

Managing cross-pollination

Onions are in the *Allium cepa* species just like shallots. Some green onions are in the *Allium cepa* species (which cross with onions) and others are *Allium fistolusum* (which don't cross with onions).

Onions are crossers. Grow them 1,000 feet from other *Allium cepa* crops in flower to avoid most crossing.

Seed yield

Seed Crop Bed Length	Imperial Yield	Metric Yield	Seed Yield
100 bed-feet (30 bed m)	4 to 7 lbs	1.75 to 3.25 kg	360K to 660K seeds
1 bed-foot	0.6 to 1.1 oz	18 to 32 g	3.6K to 6.6K seeds
1 bed m	2 to 3.6 oz	59 to 105 g	12K to 22K seeds

Growing the seed crop

Grow first-year onion bulbs for seed the same way you grow a vegetable crop. Choose your favorite onions to keep for seed. You can harvest them the same way you normally harvest onions.

You can store onions in a cold room at 32–40°F (0–4°C) with other winter crops but you can also store them at higher temperatures. As long as onions get eight to ten weeks of temperatures under 54°F (12°C) they will flower when you plant them out.

When the ground is dry enough to work, you can plant onions directly in the field. Plant onions with three rows per bed and 6 inches to 12 inches between plants. Onions stored at warmer temperatures during the winter will sprout sooner. Second-year onions start by putting on new leaves and then after a few weeks they will grow flower stalks. Stake and trellis your flowering onions to keep the stems from breaking.

Onions will take about three to five months to get to seed once they are replanted.

Fig. 7.9: *Replant your onions in a seed bed.*

Seed harvest

Keep an eye on the onion flowers—when you start to see black seeds poking through some of the flowers, you can start harvesting.

Fig. 7.10: *Trellis your onions to make sure they stand up. You can use Hortonova flower netting to support your onions.*

Fig. 7.11: *When you start to see black seeds poking through, you can harvest your onion umbels.*

Use secateurs to cut the stem about six to eight inches beneath the onion umbel. Place the stems in a bin in the same direction. Bring the stems to a dry location and place the stems on a tarp being careful to preserve the umbel orientation. Let the onion seed keep maturing on these stems for seven to ten days. You can flip the stems every day for the first few days to help control humidity.

Seed extraction

As the onion seed matures, a lot of the seed will shatter on its own. This seed is fairly clean and could be put into storage with little extra work.

You can shake the umbels to get more seed to shatter or gently rub the umbels on a screen made from hardware cloth over a bin to shatter the seeds. The more you manipulate the dry flowers, the more flower pieces will get mixed into your seed lot and the more cleaning you'll have to do.

Seed cleaning

Use dry seed techniques from Chapter 8 to get your onion seed clean. These should get your seed clean enough.

If there are still lingering bits in your seed, you can place the seed in a container with water and let the light bits float to the top. Quickly pour off the floaters. Remove your seed from the water ASAP and spread it out to dry. Use a fan to speed up drying.

Fig. 7.12: *Keep your harvested umbels all in the same orientation.*

Fig. 7.13: *As the umbels dry, some seed will shatter readily and some will stay in the umbels.*

BEETS

Beets are not the easiest seed crop to grow but they are still a very good seed crop for market growers.

The difficulty with beets is to know when to harvest the seed. If you're in a drier climate, you can leave those beets longer in the field. In a wet climate, you want to bring in that beet seed before it's been subjected to too much rain and that might mean your seed doesn't mature as well. My early beet seed crops often germinated around 45 percent. And I'm worried that if you take beets on as a first seed crop, you'll be faced with a similar challenge.

That being said, it still might be worth it for you because beets produce a lot of seed per plant. You can get a couple of pounds of beet seeds from only ten bed-feet of seed crop. With the cost of some beet varieties, a couple of pounds of seed might be good savings.

And if you have low germination seed, you can simply sow your seed thicker!

If you do choose to go this route be aware that, though ten bed feet of seed beets for a couple of generations will be fine for your variety's performance, you're going to run into trouble with having a limited gene pool if you don't buy in new stock seed every few generations. More of that in Chapter 9 in the section on population numbers.

Seed Beets at Tourne-Sol

When we want to grow a beet seed crop, we choose our favorite beets from a fall harvest of roots for our vegetable baskets. However, beet seed production is still unreliable for us and we can't consistently get high seed germination. We mostly buy beet seed from seed growers who are in drier climates than ours.

OP availability

There are a lot of OPs and F1 beet varieties on the market. The F1s tend to perform better on the edge of the season when it's colder and conditions aren't as optimal. But OP beets perform quite well.

Managing cross-pollination

Beets are crossers. They are in the *Beta vulgaris* species and will readily cross with other crops of the same species, such as Swiss Chard. Follow the guidelines in the Swiss Chard profile on page 104.

Seed yield

Seed Crop Bed Length	Imperial Yield	Metric Yield	Seed Yield
100 bed-feet (30 bed m)	15 to 25 lbs	6.75 to 11.25 kg	336K to 574K seeds
1 bed-foot	2.4 to 4.1 oz	68 to 114 g	3.4K to 5.7K seeds
1 bed m	7.9 to 13.4 oz	223 to 374 g	11K to 19K seeds

Growing the seed crop

Beets are biennials. You can grow your year-one beets with your market crop. Choose a mid-summer planting to select your seed roots. In the fall, choose the best roots for storage and leave some of the greens on the plant to keep from damaging the growing point. Overwinter your beets in a cold room at 32–40°F (0–4°C), or if your climate permits it you can leave them in the ground outdoors.

In year two, if you overwintered your beets in a cold room, plant them into soil in containers in the nursery and watch them break dormancy. Discard any roots that have not produced good leaves after one to two weeks. Plant these beets in the field with one row per bed and one and a half feet between plants.

Beet plants are quite big and sprawly. In wet climates you should trellis your plants to keep them off the ground. This can also help manage space well and keep the plants from taking over neighboring beds.

Beets will take about three to four months to get to seed once they are replanted.

Seed harvest, extraction, and seed cleaning

Because beets are chard and chard is beets, follow the directions on page 105 for chard to harvest and clean your beet seed.

CARROTS

Carrots are an important crop on market farms and you probably use a lot of seed. It would be great if you could grow that seed yourself but this is a seed crop that I would only recommend to take on when you have a bit more seed work skills.

One thing that makes growing carrot seed tricky is Queen Anne's Lace. These are wild carrots that are in the same species as your farm carrots. They are present in the countryside in many areas and will happily cross up your carrot seed plot.

Carrot seed also much prefers dry climates and it can be hard to get high quality seed in wet climates.

And carrots suffer from inbreeding depression quicker than many other crops do. This means you need to have a lot of plants to go to seed to maintain good genetic diversity (we'll talk more about this in Chapter 9 in the population size section).

All these things make carrots a better seed crop choice once you've developed your skills with some simpler crops.

OP Availability

Most of the orange market carrots out there are hybrids. And this is one vegetable class where I've found that the F1 varieties stand a few steps above OPs.

There are still some decent OPs out there, especially in non-orange carrots. But you should trial them along with F1 carrots to see what you like.

Seed Carrots at Tourne-Sol

A lot of growers in wet climates have trouble getting good quality carrot seed. Our farm has been no exception. We get our carrot seed from growers in dryer climates.

Managing cross-pollination

Carrots are crossers. They are in the *Daucus carota* species. Be aware that many areas also have Queen Anne's Lace—a wild carrot relative that is also a *Daucus carota*. They will readily cross with your carrot varieties.

In an area without Queen Anne's Lace, 1,000 feet between carrot varieties will result in very little crossing. If you have a lot of Queen Anne's Lace in your area, I would recommend you don't grow carrot seed in your market garden.

Seed yield

Seed Crop Bed Length	Imperial Yield	Metric Yield	Seed Yield
100 bed-feet (30 bed m)	2 to 4 lbs	1 to 1.75 kg	420K to 840K seeds
1 bed-foot	0.3 to 0.6 oz	9 to 18 g	4.2K to 8.4K seeds
1 bed m	1 to 2 oz	30 to 59 g	14K to 28K seeds

Growing the seed crop

Carrots are biennials. You can grow your year-one carrots with your market crop. Choose a mid-summer planting to select your seed roots. Leave some greens on the plant to keep from damaging the growing point. Overwinter your carrots in a cold room at 32–40°F (0–4°C). In some climates you can overwinter your carrots in the field.

In year two, plant your carrots in the field with two rows per bed and one foot between plants. You can trellis the plants to keep them under control. Carrots will take about three to four months to get to seed once they are replanted.

Seed harvest

Carrots produce a series of umbels. They start with one big umbel that will have the biggest seed. You can clip this when the seeds change from green to brown. You can also harvest the secondary umbels for decent-sized seed. Many growers don't try to get the third umbels and larger since the seed will be smaller and has less chance of being fully mature.

Seed extraction

Put on gloves and rub the carrot seeds to separate them. Be careful because carrot seeds can be quite sharp.

Seed cleaning

Use dry seed techniques from Chapter 8 to get your carrot seed clean. One challenge with carrot seeds is that you can wind up with many small stem pieces that are roughly the same size as your own seed. This is not a big deal if you're using this seed for your own use.

GROWING OTHER ROOT CROPS FOR SEED

The important thing about root crops is that they are almost all biennials and they need to go through a vernalization period where they get cold enough for them to flower in their second season. The requirements vary for each crop but as a general rule John Navazio recommends keeping biennial roots below 50°F (10°C) for at least eight to ten weeks.

Whether you're growing parsnips, scorzonera, or celeriac, you can use the general techniques from these different crop profiles to grow other root crops for seed. Plant them out with a bit more space in their second year to make sure they have room to grow.

Part 3

Push Your Seed Work Further

It is now time to let your First Seed Mindset go as you start to look further in what you can do with your seed work. Up until now we've focused on the idea that good enough is good enough to get you into growing seed, but these next chapters are each in their own way about effectiveness and efficiency and making a plan and carrying it out.

Chapter 8

Improve Your Seed Work with Better Tools and Techniques

THIS SECTION is about working with better tools and techniques. What you'll quickly realize is that good tools don't have to be expensive tools (though there are definitely some expensive tools out there), and that technique is even more important.

In this chapter, you'll take the time to organize your work areas and then jump into the work. I've broken the actual seed work into five categories:

- Harvesting your seeds from the field.
- Extracting your seeds from the rest of the plant.
- Cleaning your seeds.
- Testing your seeds to see how they germinate.
- Storing your seeds.

GETTING ORGANIZED FOR SEED WORK

Set Up Your Seed HQ

This is where you will do your seed work. It's where you can bring in piles of plants to dry before you extract the seed. It's where you can do some of your seed cleaning out of the wind. It's where you can leave seed partially cleaned while you wait to get back.

One place that works on a lot of farms is a nursery greenhouse. This is where you might start all your seedlings to plant out. In the early spring it is probably packed with plants, but come summer you likely only have a small portion in use for your successions. The rest of the space is great for seed work.

If you don't have a nursery you can repurpose during the summer, you could put up a caterpillar tunnel to use as a seed HQ. You can also use an outbuilding or a workshop or a garage.

The important thing for your seed HQ is that it is dry and well ventilated. It's also good if it is well lit for you to see what you're doing. It would also be great if you can keep rodents out of the structure but that is often impossible. The next best thing is to set up traps to keep rodents at bay. You can also cover some seed lots or small piles of drying plants with hardware cloth or window screen. This can also help against birds that come around to see what's going on.

Set Up a Seed Lot Tracking System

If you only ever grow one or two crops to seed each year, you can probably get away with simply labeling an envelope with the variety name and harvest year. But if you start growing many more seed crops per year, you might need a better system to track what you've grown. If you are certified organic and start selling seed, then you will definitely need to set up something to keep your certifier happy.

The easiest solution is to assign lot numbers to each seed crop you grow.

What is a seed lot?

A seed lot is a batch of seed that you store together and treat in a similar manner.

If you have multiple harvests from the same variety produced in the same year that are treated roughly the same way, then you can mix them all together to form one lot. It is much easier to handle one lot in a year for germination tests and storage, than a bunch of smaller bags that are essentially the same.

The only times I wouldn't mix the different seed harvests in the same year is if there is something particular about one of the lot. Maybe one tomato seed harvest was left too long to ferment and I'm worried it already started germinating. I would keep this harvest separate from the other harvest initially and I would do a separate germination test for this harvest. If it germinates at a similar rate to the rest of the seed lot, then I would likely mix that seed back in with the others.

Another case where I might keep seeds separate is if one batch was significantly more dirty than the other, or perhaps the seed got wet, or something else that might reduce the quality or make it harder to work with. In this case you could create a separate lot for the seed, and put a note on this seed why it might cause problems later on. You then have this seed as a backup in case you ever need it.

The anatomy of a seed lot

What I like to do is have a five-digit code where the first two digits are the last two digits of the year and the following three digits correspond to the different seed crops produced that year. So you might have four seed lots on your farm:

23001, 23002, 23003, 23004. These were all produced in 2023. 23001 would be the first lot that year, 23002 the second and so forth. These numbers give you the opportunity to go all the way up to 999 varieties in a given year. This may be way more than you ever think you'll use. But it's good to give yourself room to grow.

On our farm, we also buy seed from other growers. With three digits, we do have more than enough numbers to fit those numbers into the lot numbers for a year. But instead we add two letters to the beginning of the lot number to indicate where the seed is from. For the seed we grow on our farm we use TS (for Tourne-Sol). We have a series of other two-letter codes for each of our seed suppliers.

Having the source and year as part of a seed lot means you can quickly pull together a lot of information from the lot. Earlier this year, Amiya, who is in charge of direct seeding at Tourne-Sol, mentioned that some of the bok choy hadn't germinated well. I asked for the seed lot. She told me it started with TS17. I immediately knew we grew that seed ourselves six years ago. So it wasn't surprising that the germination had dropped after six years. Amiya mentioned that this same lot had germinated well the previous year. I loved hearing that because it had been doing well for five years.

When do you assign seed lots?

You can assign lots to your seed crops as they get harvested. But I like to assign them right from the moment I'm crop planning. That way I don't accidentally assign two crops the same lot. We also use the lot number to keep track of crops at every stage from planting through seed cleaning.

Label Your Seeds at Every Step of Their Journey

Here is the most important advice in this book: Label everything always.

In the field, crops might be easy to distinguish, but once plant parts dry down and seed gets extracted, it gets pretty hard to distinguish many seed varieties one from the other. (Bok choy and rapini and turnip seed all look the same.)

There are three phases where you can label a crop:

1. When a crop is growing in the field.
2. When a crop is harvested or in post-harvest cleaning/drying.
3. When seed is in storage.

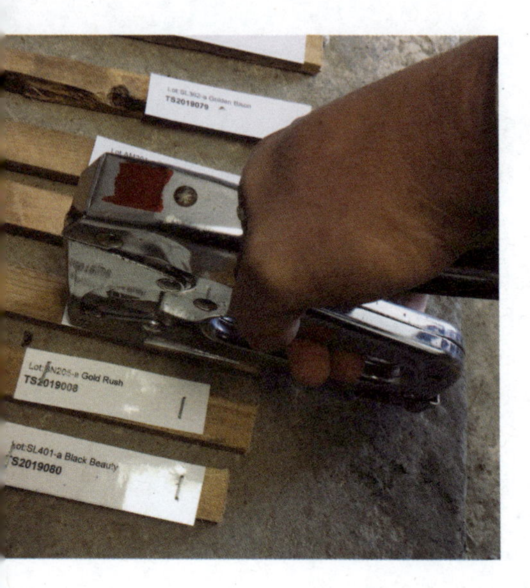

Fig. 8.1: *These are the laminated tags we use in the field. On the bottom tag you can see the product code (SL401-a), the variety name (Black Beauty) and the lot number (TS2019080).*

Label crops growing in the field

You can identify crops in the field using wooden stakes. You can use a sharpie to write on the stake or print information on a nursery label that you staple to the stake.

I like to laminate a printed tag and staple that on the stake.

The variety name is the main thing to have on your field label. If you use seed lots and have a lot assigned to the crop, put that on it too. If you have a code or SKU for the crop, that should be there also.

The easier it is to read the label the better!

Label your seeds during harvest, cleaning, and drying

When you harvest a seed crop, you want to identify that seed immediately. At the least you can use some masking tape to write down the variety name/SKU/lot from the field label. You could also add your harvest date.

We have pre-made forms that we fill out and clip to each seed lot as it comes out of the field. These labels are then moved along with the crop as it is threshed and then transferred to buckets as the amount of material gets smaller through cleaning. We often also add masking tape as a back up. There is a very specific sinking feeling when you find an unlabeled seed lot in your seed HQ and it looks like four other things you grew this year. It is not a good feeling.

Clothespins are great tools to clip labels to buckets and tarps to keep track of your seed. They are easy to clip and unclip as your pile of seed and chaff moves through seed cleaning and usually winds up in smaller containers.

If your label is big enough, you can also use it to keep track of cleaning notes and cleaning dates. You can also use it to record any selection criteria you might have done when you harvested these plants.

Do be careful though that your labels don't sunbleach.

Labeling your seeds in storage

You need two labels in storage: one label written on the outside of your container and another label on a piece of paper or kraft envelope inside the container.

Double labeling means you don't have to worry if the outside label gets worn off. It also means that you have a way to catch it if the seed winds up in the wrong container.

Fig. 8.2: *Label the outside of your storage containers.*

Fig. 8.3: *And put a label inside your storage containers too!*

(There is another sinking feeling that comes when you discover that the labels *on* the container and *in* the container don't match.)

HARVESTING YOUR SEEDS FROM THE FIELD

One thing about most seed crops is that their seed usually matures over a period of time on the crop in the field so there is no one time when all the seed is ready.

This leaves you with two main approaches to harvesting seed crops:

1. Go in one to two times a week and harvest what is ready.
2. Or harvest the whole plant when half to three-quarters of the seed is ready and accept that you will not get every seed.

Regular Seed Harvests from the Same Plants

With this approach you keep an eye on your seed plants to see how mature the seed is. When you see that there is mature seed on the plant, you go and harvest the mature fruit or dried seed heads or dried pods. In some cases you might be shaking the actual seed off the plant.

You then bring your daily harvest to your seed HQ and for many crops you can just add the harvest to a tarp beside your previous harvest.

Harvesting from the same plants regularly usually results in the highest quality seed because the seed has fully matured on the plant. It also means that you usually get the highest yield from your plants because you harvest the seed promptly before it shatters.

Fig. 8.4: *Lettuce is a crop that can be harvested multiple times. Each time, carefully shake the flowers into a bin.*

There are some crops where this is the only way to make sure you're harvesting quality seed.

This is the case for many flowers where harvesting too much plant material can make them difficult to clean. It's also the case with fruiting vegetables like tomatoes and peppers that keep maturing over a long time.

But it can be hard to stay on top of regular seed harvests in the middle of the summer when all your market crops are also coming in. This approach works well if you only need small amounts of seed for your own farm use. But for some crops this is too much work to get bulk seed.

Harvest the Whole Plant

With many crops, you can skip all those visits to pick small amounts of seed and just pull the whole plant out of the ground when most of the seed is ready. This works really well with brassicas, beans, peas, and other crops where the seed doesn't shatter easily.

After you've pulled the plant out of the ground you can then spread it out on a piece of landscape fabric in a dry area. If you are in a dry climate, you might be able to let the plant dry down outside. In a wetter climate, you should bring your plants into your seed HQ. If the plants still have green on them, it is important to flip the plant daily to keep the pile ventilated to avoid composting or rotting.

Generally, you can wait until two-thirds or more of the seeds are mature on the plant to cut it down. You should always cut the plant stem above ground and leave the roots behind. This keeps the dirt on the roots in the field rather than accidentally mixing with the seed.

Sometimes the weather might mean you want to harvest your plants when the seed isn't quite as mature. If you know there is an intense storm coming or a very long period of wet weather, you should consider bringing your seed crop in early. Very wet and windy weather can shatter seed on the ground and knock your plants over, putting seed heads in contact with the ground where they can start to germinate or simply rot.

Early frosts can also be a reason to bring your seed crops in early. Though mature dried seed can freeze without problem, frost can damage the immature seed on frost-sensitive plants. Bringing plants in early can be stressful for farmers but you would be surprised how much some crops can keep maturing even if the plant is not in the ground. And even if an early harvest reduces the yield, the seed quality will be much better than if you'd left the seeds out in bad weather.

For bulk seed crops you'll find that whole plant harvests are much quicker than regular small seed harvests. One word of advice—try to keep all your plants in the same orientation during harvest and when you spread them out afterwards. This will make seed extraction a lot easier.

Fig. 8.5: *These whole bean plants have been harvested with the pods still on the plant. They are drying down in a greenhouse.*

When is it Time to Stop Harvesting Seeds?

If you are harvesting some crops one to two times a week, you'll notice that for some of them there starts to be fewer and fewer seed heads or fruit on the plant. For crops like these, it's pretty obvious when to stop harvesting. But for other crops the plants keep putting out flowers or fruits. There was a cherry tomato last year that produced twelve five-gallon buckets of cherry tomatoes every week for six weeks. Each harvest gave a lot of seed, way more than we'd be able to sell in many years. For crops like this it's important to know when you have enough seed.

In Chapter 10 you will estimate how much seed you need for each crop. If you're dealing with a pound or two of seed and harvest and cleaning is quick, then it's fine if you harvest twice as much as you need. But if it takes a lot of time, you shouldn't be harvesting tons more than you need because that's time you could be doing something else. If you sell or share seed, you might decide to harvest more on the off-chance that someone wants it.

Always remember you can stop harvesting seed whenever you have as much as you want. At that point till or mow the planting to keep it under control.

Harvest Tools of Choice

Do you need a combine? No, you don't.

For market growers, seed harvest is rarely the bottleneck in your seed work. Cutting and putting crops on tarps can go pretty quickly. We harvest an acre

Fig. 8.6: *Our trusty harvest sickles.*

of seed pretty much by hand as full plants or regular harvests. I've seen farmers with seven to ten acres of seed mostly harvesting by hand.

The two seed harvest tools that we love on our farm are harvest sickles and pieces of landscape fabric.

We use Zenport harvest sickles. I saw mention of these in a Wild Garden seed catalog and then found some online. I'm not sure that these are the same exact models that the Wild Garden seed crew use. These harvest sickles let you work in small areas and cut the stems with a quick pull. They don't require a big whack to get through the crop.

In my early seed harvest days, I would harvest into bins and then transfer onto tarps in my seed HQ. These days, we mostly harvest directly onto pieces of landscape fabric and then roll up the pieces with the seed in the middle. We pile up the rolls on a wagon and then unload and unroll the pieces in our seed HQ. We buy four-foot-wide rolls of landscape fabric and cut them into ten-foot lengths using a propane torch to burn through the plastic. The torch melts any loose strands and leaves us with a clean edge that won't unravel.

Fig. 8.7: *Radish plants rolled up in landscape fabric about to head back to seed HQ.*

EXTRACTING YOUR SEEDS FROM THE REST OF THE PLANT

There are two very different ways to get seed out of a plant: wet-seed extraction, which can use water, and dry seed extraction, which avoids water at all cost.

Wet-seed Extraction

Wet-seeded crops are grown in fruits with wet fleshy insides, such as tomatoes These seeds come out of the fruit already a bit humid. Because they are already wet, it's safe to use water to clean these seeds.

The basic approach to wet-seed extraction is to cut the fruit open without slicing through the seed, scoop out the seed and some of the flesh, and then decant the seed to remove light bits.

To decant seed, you put the seed and attached bits into a container. Fill up the container with water; the seed sinks to the bottom, and the lighter stuff floats to the top. You can then pour the lighter stuff off. Add more water and repeat the process until the seeds are clean. This works for most wet seeds with a couple of exceptions.

Fig. 8.8: *Decanting a bin of cucumber seeds. You can see the light immature seeds floating off.*

Wet seed challenges

For most seed crops, good seed sinks and bad seed floats. For these crops, the light seed will pour off from the mixture. However, there are some crops where good seed will float. Specifically, squash. Go to the squash section to read more about what to do in this case (page 59).

Some crops also have gelatinous sacs around the seed. You will need to ferment these seeds to get rid of these membranes and make it easy to clean these seeds. This is the case for cucumbers and tomatoes. Read their profiles for more on those steps.

Get your wet seeds dry ASAP

Once you've extracted a wet-seeded crop from the fruit and rinsed it, it is now susceptible to germination if it stays wet too long. You want your seed to get pretty dry in a few hours. You should spread the seed out on screens or trays and put fans on the seed to improve air circulation. You can also put your seeds in jewel bags or machine-part bags and squeeze some of the moisture out. You can leave those bags to dry in the sun in a greenhouse with fans on them or you can build a forced-air dryer to speed up drying.

Wet seed extraction tools for bulk seed

Millet Wet Seed Extractor: This machine crushes up fruit and then sprays it out over a screen. The wet seed falls through the screen into a bucket and the fruit flesh and screen shake into a separate bin. We use the Millet for tomatoes without adding any additional water. The collected tomato seeds and juice are then fermented. We add water to the Millet when we're extracting eggplants and peppers. We decant the pepper and eggplant seed bins and then collect the seed to dry. Theoretically, you can clean cucumber seeds with this machine, but I find it jams too much. The Millet is a great machine if you're producing many lots of one or two pounds of tomato or pepper seed in a year.

Fig. 8.9: *You can feed tomatoes or peppers into the hopper on the top of the Millet.*

Fig. 8.10: *The bin on the far left is full of the pepper flesh. The bin on the right is full of seeds and water.*

Cucurbit Extractor: This machine squishes cucurbits of all kinds and separates the seed from the flesh. They are probably too expensive to justify purchasing for smaller cucurbit lots, but should be considered if you're producing acres of cucurbit seed.

Dry Seed Extraction

Dry-seeded crops are those where the flowers and pods of the plant are fully dry at maturity. You should avoid adding any water to these crops during seed cleaning because you can create the right conditions for germination.

One of the secrets to working with dry seed crops is to wait until the crops have had enough time to dry out before trying to extract the seed. And make sure to work on a dry day or during a dry part of the day. It is amazing how much ambient humidity can get picked up by dry plant material. Any amount of moisture can keep seeds from separating from their pods or other plant material.

Dry seed techniques

There are a range of ways to extract dry seeds from the plant:

Extract by hand: If you have a pile of pods or flowers, you can just sit down with them and open them up and take out the seed. Drop the seed in one bowl and put the rest in another bowl for compost.

With some crops you will wind up with very clean seed this way. This would be the case for hand shelling beans and peas. But it's also true for strawflowers or nigella. There isn't much extra cleaning you'd have to do afterwards.

With some crops, you can remove some of the dry flower parts and wind up with a pile of seed with some similar-sized chaff. This would be the case for zinnia.

Though you can technically extract any seed by hand, this works best for crops where there is a pod or seed head that is convenient to harvest on its own and then to handle on its own. This might not be great with brassica siliques since they are very small to handle individually and are likely to shatter.

Collect shattered seed off a tarp: For many seed crops, you wind up spreading your harvested plants or stems on a piece of landscape fabric to continue drying. During this time some crops will release their seed as it matures. The seed falls on the tarp.

As you remove the dry plant material you'll find a pile of seed lying under the plants. This would be the case for onions and other allium seeds. You can scoop or blow off some of the chaff and wind up with some pretty clean seed. This seed is often the highest quality seed of the harvest. There might still be seed that hasn't shattered off these plants, but it might not be as mature as the seed that has shattered. And you will need one of the following methods to extract the seed, which will result in seed that is a lot less clean then what you just picked up off the tarp.

Shake seeds off plants: A variant of the shattered seed approach is to take the plants and shake them gently into a bin or a bag. This technique will loosen any mature seed and help it to shatter off the plant if it is still stuck on it. You don't want to use too much force in your shaking, so that only mature seed shakes off and immature seed doesn't. If you're gentle with your shaking, your seed will be easy to clean. If you're too rough, you can wind up with a lot of small bits of stem or other plant parts mixed in with your seed, in addition to a lot of immature seed. This will be a harder seed lot to clean.

Knock seeds off with a stick:. This is pretty much what it sounds like. This is where it is especially great to have all your drying plants lying in the same orientation. It makes it easy to go down the row and tap the seed heads to let them shatter off the plants. This is great for something like dill.

Dance seeds off their plants: Move your tarp of dry seeds to the ground and check your boots to make sure the bottoms are clean. Then get on the plants and start dancing to shatter the seeds off the plants. This works great with brassicas. This is also what you would do if you harvested whole plants of beans or peas, though they can require a little more force. After dancing, pick up the stomped plants and shake them a touch and transfer them to a pile destined for the compost. Collect what's left on the tarp and put it in a bin. If you have screens in the field, you can screen this stuff to remove some of the chaff. You can also winnow some of this seed in the wind to reduce the amount of material you're bringing back into seed HQ. If you have many piles of plants, you can instead drive over the plants with a pickup truck or a tractor to shatter the pods.

Dry seed extraction tools for bulk seed

Converted Wood Chipper: Many seed growers convert wood chippers into seed threshers. This involves modifying the hammers on the chipper and slowing down the speed of the machine. I've also seen a number of growers use unconverted wood chippers to thresh seed plants. If you go this route, try small batches first to see what the results are like.

Reid Allaway, Tourne-Sol's resident tinkerer, converted a soil shredder in a similar way by replacing the AC motor with a DC motor and speed controller, allowing us to adjust motor speed to thresh more delicately. We've mainly used this thresher for beans and peas but we've tried many other dry-seeded crops in it too.

Fig. 8.11: *The Tourne-Sol soil shredder taking a break from threshing some bean seeds.*

Stationary Threshers: There are a range of stationary threshers out there for a range of prices. Many of them might be too expensive if you're not growing a lot of seed.

One thresher that we've found at a more affordable price range is the MT-860 multi-crop thresher. We've been able to handle a wide range of crops with it including peas, okra, radish, and spinach. It has really sped up extracting these seeds.

One crop it has not been successful for yet has been beans. We've been disappointed with the number of split beans that have come out of the machine. We still use our modified soil shredder for beans.

Fig. 8.12: *Threshing sorghum with the MT-860 multi-crop thresher. Ear protection and dust masks make this job much more enjoyable.*

Cleaning Your Seeds

Whether you're extracting wet-seeded crops or dry-seeded crops, at the end you'll wind up with a pile of seed mixed with a little bit to a lot of chaff. That's where seed cleaning comes in.

There are two main techniques to clean seeds: winnowing (using wind to remove light chaff) and screening (using size to separate your seed from differently sized objects.) Seed cleaning comes down to alternating winnowing and screening until you get your seed clean enough.

There is one more technique to deal with whatever bits remain after you've gone as far as winnowing and screening will take you: hand picking.

I mentioned that the secret to dry seed extraction is to always work with dry material on a dry day. That secret works here too. But an even more important secret is: **"Don't try to save every last seed when you're cleaning seed."** With most seed lots, it is pretty easy to get the first 80–90 percent of the lot pretty clean. The last 10–20 percent can take a lot more time to get clean.

If you're struggling with that last ten percent, put it in a separate container and label it indicating that the seed is not quite clean. You can come back to it later if you need it. It's usually a better use of your time to move on to other farm tasks than struggle with the itty bitties left in some of your seed.

Winnowing Your Seed

Winnowing is the process of pouring seed into the wind or blowing the light chaff away. I'm amazed how many new seed cleaners have not tried to winnow their seed. I'm not sure if it's a fear of losing seed in the wind or if setting up fans seems like too much work. What they don't know is that winnowing is one of the most satisfying parts of growing seed. If you've never winnowed your seed, it is high time you do!

Fig. 8.13: *It's easy to do a first winnow outdoors in the wind when you're working with heavy seeds like beans.*

Winnowing outside in the wind

You can use the wind outside to winnow your seeds. Put some dirty seed in one container and pour it into another container. If there is a breeze, you'll see it blow the lighter stuff into a separate stream from the heavier seed.

Natural wind is tricky because it is not always there when you want it. And sometimes there is way more than you want. Stronger winds are usually fine for big seeds like beans, peas, and grains. But stronger winds can be tricky to manage for lighter seeds.

You can winnow smaller seeds more safely by reducing the height from which the seed drops. Simply pouring seed from a container in one hand to a container in another hand can be enough to winnow light seed.

When you're new to winnowing, you should work over a tarp to capture anything that falls from unexpected gusts of wind. This isn't bad advice for more experienced winnowers either.

Winnow with box fans

If you want to get out of the unpredictable outside wind, you can create your own wind with an electric fan.

Fig. 8.14: *Set up two fans in your winnowing station to get more wind control.*

Fig. 8.15: *Winnowing with two fans. The cleanest seed is in the bin on the right.*

Any fan can do but square box fans are a lot easier to control. Two box fans placed one after the other will help even out the airflow. It also gives you a few different speed options.

When you're winnowing with fans, place the fans on a table and set up three bins on a tarp to capture your seeds. The goal is for the heaviest seed to fall into the first bin and the other two bins to capture different grades of chaff. You can check the other bins to see if you're losing any seed in the other bins and adjust your speed accordingly.

Screening Your Seed

There are two ways to screen your seed. You can use screens that are bigger than your seed. This removes larger pieces of chaff. You can also use screens where the holes are smaller than your seed to remove the smallest stuff from your seed lot. Generally you can winnow out the stuff that is smaller than your seed because it is usually lighter than your seed. Sometimes you have small dirt pieces or other material that is almost the same weight as your seed and it resists winnowing. This is where you want finer screens to work with.

Fig. 8.16: *A collection of colanders and strainers for cleaning seeds at Hawthorn Farm Organic Seeds.*

Kitchen colanders and spaghetti strainers

The first screens you should try are whatever colanders you have in your kitchen. You can also raid your local dollar store and second-hand store. It shouldn't be hard to find colanders with a few different mesh sizes.

Even after we got professional seed screens, I still had a few favorite colanders that I loved for cleaning seed. It was a sad day when my red colander cracked beyond repair—it was my favorite screen for radish seeds.

Homemade screens

The second screens you should acquire are ones you make yourself using hardware cloth from a hardware store.

Hardware cloth comes in different sizes but ⅛", ¼", and ½" are pretty common. These also give you a nice range of sizes to accommodate a lot of seed sizes.

Fig. 8.17: *This screen is made from ¼" hardware cloth.*

Make a frame for your seed using pieces of 1×2-inch wood. You can staple, screw, or nail the cloth to the frame. You can also pinch the cloth between two pieces of wood.

You can place your 1×2s flat which leaves a small ledge to keep seed from running off. Or you can put 1×3s or 1×4-inch perpendicular around the edge to create an open box that will hold seed and chaff.

You can make screens any size you want. Twelve-inch by twelve-inch is fine for small lots. I like to make screens that are the same size as the Rubbermaid bins I use for seed cleaning.

You can even make bigger screens and add legs like a table and use it to screen big batches of seed.

Fig. 8.18: *We purchased these screens and then made the frames ourselves.*

Purchased seed screens

You can also order professional seed screens. These come in a few different shapes: Round holes, long slots, herringbone openings, or wire mesh rectangles. And you can get all these shapes in different sizes. Professional seed screens can cost over $50 a screen. It is an investment to get enough sizes to cover a range of different crops. But having the right size screen makes a big difference getting the seed to the last step of cleanliness. If you start dealing with larger volumes of seed you should go ahead and invest in some seed screens.

Handpicking Your Seed

No matter how good you get at winnowing and screening, you won't be able to get all the unwanted bits out of all your seed lots—split beans that don't winnow out, blackened pepper seeds, sunflowers with holes. At that point the solution is to spread out your seeds on a cookie tray and use your fingers or tweezers to pick the bits out of your seed.

Do you need to do this?

If this seed is for your own farm, probably not. You can adjust your seeding rates as a function of this stuff. If you're selling the seed, you should probably get most of the bits out of your seed.

Fig. 8.19: *Picking out split beans, rocks, and other things from bean seed.*

If you do wind up handpicking a seed lot, you should think about why there are these bits in the first place. There might be an earlier step in your seed work where a little more care might reduce the need for picking. If you have black seeds in your peppers, this probably comes from bulk squishing to extract the seeds. Taking more care to look at the fruit that is getting squished and removing anything that shows damage will probably reduce the rotten seeds in the lot. Or simply processing peppers more quickly and not letting them after-ripen as much would do a similar job. If you have a lot of split beans, threshing the plants when they are a bit less dry might keep more beans from splitting.

But it is good to remember that hand picking the bits out of seed is an option. Turn on the radio or your favorite podcast and get picking!

Secondary Seed-Cleaning Tools

Just as in any field, there are specialized expensive seed-cleaning tools out there. They can definitely improve your seed cleaning. But do you really need them?

These tools do replace winnowing and screening your seed as I've described above. You should always start by removing whatever stuff you can with screens and fans. This will give you some pretty clean seed to put into the fancier tools below. Then you can focus on getting some of the harder-to-remove things.

Be aware that you still might have some handpicking to do after these steps too!

If you're only growing a few seed lots a year, the tools below are probably overkill. If you're growing many seed lots, especially if you're selling a lot of seed, these tools might be worth the investment. Or even better, get together with some other local seed growers and pitch in to purchase this equipment and have it travel between farms.

Fig. 8.20: *The seed spiral separator.*

Spiral separator

This tool separates spherical things from non-spherical things. Pour your seed into the hopper at the top and watch the seed go down the spiral track. Any spherical seed will accelerate more than non-spherical seed and jump off the spiral track into a separate chamber and empty out from a different chute. The spiral separator works really well with peas and brassica seeds.

My favorite thing about the spiral separator is that there are no moving mechanical pieces. So there is no maintenance other than cleaning out the machine with an air compressor between lots.

Office Clipper fanning mill

The Office Clipper integrates both winnowing and screening into one fancy machine with many moving parts. Seed falls through a top screen that removes things larger than the seed. It then passes over a bottom screen that removes things smaller than the seed. It then falls down a column in front of a fan and blows out anything lighter than the seed.

The Office Clipper is a great machine but I would only recommend it to folks running seed companies who are trying to get their seeds perfect. Most seed growers do not need this machine.

Air columns

Air columns separate seeds by density. Seed is put into a chamber and air blows through the

Fig. 8.21: *The Office Clipper fanning mill.*

seed lifting most of it into the air. The lightest material gets sucked into a separate area and the heaviest seed falls back down.

Air columns are great to separate seed or material that is a very similar shape to your seed but just a bit lighter. This is stuff that could possibly get winnowed out with a fan, but an air column provides a lot more precision.

There are a number of different air columns out there, including quite a few DIY versions. Some work by putting batches of seed into a chamber and others you can just pour seeds through the machine.

Real Seeds DIY Seed Cleaner

This is a type of air column. It was developed by the Real Seed Company in Wales. They have shared their plans and it is fairly straightforward to use. I know many seed growers who've built them and they all love them.

Winnow Wizard

The Winnow Wizard was developed by Mark Luterra while he worked at Wild Garden Seed. This tool replicates the motion of winnowing with a fan but in a more reliable way. You put the seed into a hopper at the top of the machine and the seed flows into a thin stream in front of a fan. The wind from the fan

Fig. 8.22: (Left) *The South Dakota Seed Blower.*

Fig. 8.23: (Right) *Hawthorn Farm Organic Seeds built their own Real Seeds DIY Seed Cleaner.*

Fig. 8.24: *The Winnow Wizard.*

goes through a series of screens that create a laminar flow that is consistent without any gusts. Even though this is an elaborate machine, the seed doesn't actually travel through any areas that are difficult to access. This means it is very easy to clean out the Winnow Wizard between seed lots.

This is the one seed-cleaning machine that consistently comes up when I interview folks for the Seed Growers podcast. If you were going to invest in one seed-cleaning tool, this one will probably make the biggest difference to you.

Freezing Seeds for Pest Control

For most seeds, winnowing and screening will remove any insects that were mixed in the chaff, but in a few cases you might still have insect eggs in the seeds. These eggs can hatch and wreak havoc on the rest of your seed lot. This is especially a problem with weevils in beans and peas, but it can also be a problem in grains and sunflowers.

Freezing your seeds will control any insect eggs. To freeze your seed, start by putting it into a sealed pack or container. Then put your seed into a freezer for a week. If you live in a climate where it gets very cold for long periods of time, you can also just put your seed bins in an unheated outbuilding for a couple of weeks.

Make sure your seeds are very dry before freezing them. Read the bean and pea crops profiles for how to do this with those crops.

When you remove seed from the freezers, let the closed containers and seed come to room temperature for a few hours before opening them to avoid any condensation.

We systematically freeze all beans, peas, and sunflower seeds at Tourne-Sol. We only freeze other seeds if we see insect signs.

TESTING YOUR SEEDS TO SEE HOW THEY GERMINATE

You have all this clean, good looking seed, but will they germinate? The only way to know is to set up the seeds in good growing conditions and see whether they sprout.

Below I go into all the variables to set up reliable and consistent germination results, but basically you're just starting a bunch of seedlings almost the same way you would in your nursery in the spring. The only thing different is you are going to grow them very close together and not keep the seedlings once they sprout.

The goal of a germination test is generally to know what percentage of the seeds will germinate. Though precise germination rates have many benefits, you might just want a general impression of how well a seed is doing. Sprinkling seeds in potting soil, watering them, and putting them in your nursery is an easy way to have a general impression of how well a seed lot germinates.

Usually folks only do a germination test once the seed is clean, but you can actually test for germination at any point from the moment that the seed is mature. Performing a germination test at different points can give you an idea whether you should go back and winnow your seed again to get rid of lighter seed or just leave it as is.

Germination Temperatures

I used to get so frustrated with my nigella seed when it was germination test time. We would test a hundred seeds but barely any of them would sprout. Yet when we'd sow these seeds in the field, they would all germinate. One year we tested nigella germination at 60°F (15°C) instead of in a warm germination chamber and those seeds sprouted so quickly. Now we pay a lot more attention to temperature when we run germination tests.

Crops generally fall into one of three temperature groups.

- 86°F (30°C) for 8 hours and 68°F (20°C) for 16 hours
  - You will need to set up a germination chamber to monitor these temperatures.
  - If you can't alternate temperatures, aim for 77°F (25°C) for 24 hours a day.
  - Most crops fall in this category. This is my default when I'm testing the germination of a crop for the first time.

  - Crops in this temperature range include amaranth, beans, brassicas, beets, chard, basil, oregano, thyme, zinnia, sunflowers, okra, tomatoes, peppers, eggplants, squash, cucumbers, melons, carrots, and parsley.
- 68°F (20°C) all the time
  - Indoor room temperature is often good for these crops.
  - Crops in this temperature range include onions, leeks, peas, lettuce, and celery.
- 60°F (15°C) all the time
  - If you have a cold room at 60°F (15°C) where you store peppers, cucumbers, and tomatoes for market, you can also use this space for your 60°F (15°C) germination tests.
  - Crops in this temperature range include: orach, spinach, nigella, poppies, mâche.

Many seeds will still germinate if they are in the wrong temperature range. But they will take much longer to do so. If you want quick germination, get those seeds at the right temperature!

Germination Mediums

If you want to set up germination tests with what you have on hand, you can use humid paper towels or coffee filters and put them in a Ziploc. This can get the job done.

Fig. 8.25: *Tomato seeds germinating on blotter paper in a petri dish.*
Credit: Emily Board

You can also just use your nursery setup for germination tests using the same methods as if you were about to start seedlings. If your nursery is set up and you only have a few tests to do, this is a fantastic way to go.

If you want more control, you can use petri dishes with different mediums.

1. Seed is placed between two moist blotter papers.
   a. Crops: onions, leeks, brassicas, beets, chard, carrots, radishes (I know, they are brassicas!), squash, melons, watermelons.
2. Seed is planted in sand.
   a. This is good for big seeds. The sand provides some extra moisture without soaking the

seeds. We buy sandbox sand from the local hardware store for these germination tests.

b. Crops: beans, peas, corn.

3. Seeds on top of humid blotter paper.
   a. These seeds are exposed to the light, which helps these crops germinate. (Especially when they are freshly harvested.)
   b. Crops: lettuce, tomatoes, peppers, eggplant.

Other Germination Requirements

For most seeds, having the seed at the right temperature in a wet environment and possibly with some light exposure is all you need to get them to germinate. Some types of seed need additional care to be able to germinate. For example, Perilla germinates better if it is pre-chilled, and many okra varieties need their seed coat to be scarified (nicked with a razor blade or nail clippers) to be able to germinate. If you use techniques like these to start your seedlings, you should also use these techniques to germ-test your seeds.

Occasionally, you have seed that seems to be good but it just doesn't germinate. Your seed might be dormant. This can happen when lettuce seed is exposed to temperatures above 86°F (30°C) for too long. It can also happen to brassica seed when it is still freshly harvested. One way to break seed dormancy is to soak the seed in water for 24 hours while changing the water every couple of hours. Drain the seed and then set up your germination test. Dormant seed will often pop out of the ground after you do this!

Fig. 8.26: *Setting up a germ test.*
Credit: Emily Board

Set Up your Germ Test

1. **Count a certain number.** You can use whatever number you want but 100 seeds is a nice number. It's large enough to give you a good idea of the seed lot. It is also very easy to calculate a germination percentage.
2. **Label your container** before you start the test. We label all germination tests with the crop code, the lot number, and the date we started

the test. We label the blotter paper and we label the petri dish lid for the test. Double labeling keeps big mistakes from happening!

3. **Prep your medium and moisten it.** You want it to be wet enough to hold moisture for a few days but not so wet that the seeds rot.
4. **Place your seeds on or in the medium.** Try to space them out so they don't touch each other.
5. **Put your test in a germination chamber.**

Germination Chamber

You can make a germination chamber from a fridge that is not running. All you have to do is place an incandescent light bulb into the bottom of the fridge to heat the chamber up. You can also add a fluorescent light to act like UV rays for seeds that need light to germinate.

Day Phase: The light bulb is on a timer so it only turns on for eight hours. It is also on a thermostat, so that the light bulb turns off when the fridge temperature goes over 86°F (30°C).

Night Phase: Let the fridge cool naturally until it goes down to 68°F (20°C) (close to ambient temperature).

Counting and Evaluating Germination Tests

You can do this on a paper log or directly in a spreadsheet. Set up a germination log to keep track of your tests. You can get a copy of one log at sheets. seed farmerbook.com.

Table 8.1

Date Packed	Lot	SKU	Product	# seeds tested	Date 1	Count 1	Date 2	Count 2	Date 3	Count 3	Count Total	Germ Rate
2023-12-01	TS22 032	SL353	Tomato Jaune Flammée	100	2023-12-04	94	2023-12-07	1	2023-12-10	0	95	95%
2023-12-01	TS23 015	SL231	Tomato Black Cherry	100	2023-12-04	90	2023-12-07	2	2023-12-10	0	92	92%

Fig. 8.27: *Counting bean seed germinating in sand on a cafeteria tray.*

Fig. 8.28: *A tally counter helps keep track when counting germ tests.*

Every few days, take out your germination test and see what has sprouted. Count the sprouts as you remove them. Tweezers are great for grabbing individual seedlings.

A tally counter can be a great help when you're removing many seedlings from a test. Hold it in one hand and click each time you remove a seedling with the other hand. Record the number that germinated on your germ log.

To calculate your germination rate use this formula: (germinated seeds ÷ sown seeds) × 100 = germination percentage. If you sow 100 seeds then that math is pretty easy!

If you're selling seed you'll need to make sure your germination rates exceed any legal standards for your country. You get to make the call on what is acceptable when you're using the seed on-farm.

STORING YOUR SEEDS

Now that you've got clean seed and you know that it germinates well, it is time to store your seed in a way that preserves its quality.

Storage Conditions

Seed stores best in a dry area that is kept at a constant temperature. Ambient room temperatures are fine as long as they are maintained without any big variations. If temperatures go from 68°F (20°C) during the day to 50°F (10°C) at night (or

Fig. 8.29: *Storing seeds in jars at Hawthorn Farm Organic Seeds.*

Fig. 8.30: *This envelope is both a label to make sure the seeds don't get mixed up and a moisture indicator.*

even lower) regularly, your seed will think spring is coming and start to want to break dormancy. This will deplete the seed's energy reserves and the seed will not store as long as it would had it been at a constant temperature.

Cooler storage temperatures are better than room temperature as long as they are constant. You could set up a seed room with an air conditioner or CoolBot and keep it at 50–60°F (10–15°C). You can go colder than that but that isn't necessary for most farmers growing seed for their own use. And remember to keep the temperatures constant!

If your seeds are dry, you can store them in sealed containers. These might be jars, Ziploc bags, feed sacks, or other containers. The type of container doesn't make too much of a difference.

Only Store Dry Seed

You should only put dry seed into storage. Humid seed will start to mold and might even germinate.

Letting seed dry in a ventilated area for a few weeks will help get your seed dry enough for storage. You should also pack your seeds on a dry day or in a dry room to avoid ambient humidity getting trapped in your containers.

Despite your best intentions, seed that is not quite dry enough can accidentally be put into storage. You should keep an eye on your seeds. When you open a container to handle a seed lot, you should feel the inside label to see if it crinkles. If it doesn't feel crisp, your seed is too moist to be stored in a sealed container. Get it out of that container and spread it out before it gets moldy or starts to germinate.

I learned this moisture indicator trick from Frank Morton of Wild Garden Seeds. It has definitely helped me on a couple of occasions when I was too hasty in packing up seeds!

How Long do Seeds Last?

I've been surprised how long seeds can live for. For years we ran our seed business out of our apartment before we moved the seed office to the farm. During that time, we stored bags of seeds in plastic totes in a closet.

In our apartment we didn't have much sophistication in our temperature control systems—blinds on the windows and a couple of fans. These were far from ideal seed storage conditions during summer heat waves. We didn't have air conditioning, and our home and seeds were subject to the whims of the weather. Seeds were definitely dry enough but they got quite warm under a number of conditions.

In these apartment conditions, there were tomato seeds that stored for over nine years with over 90 percent germination rates. Some beans and brassica seeds kept for five or six years. Yes, there were a few lots that did not do as well. There were some beans that lost germination after two years and a few tricky pepper seed lots. But those were the exceptions—most seeds did quite well in less than ideal conditions as long as they were dry.

There are definitely some seeds that have reputations for not storing for long, such as parsnips, onions, and leeks. I have limited experience with storing parsnip seed but in regards to allium seeds, it's been my experience that they last three to four years in good conditions.

All this to say that it is definitely worth setting up good temperature control systems to store your seed, but if you aren't able to, you can make do by storing seeds in less than ideal temperature if they are dry, especially if you monitor germination rates and plan on growing new seed lots every three to four years.

Chapter 9

Become a Seed Steward

ALL THE SEEDS YOU SOW are only here due to the work of countless generations of seed keepers that came before you. As you begin to harvest seed you are joining this lineage. If you really fathom this, it is an overwhelming notion. That is why I've been encouraging you to look at one season at a time without a much larger picture. It is time to change that frame of reference.

Whereas the First Seed Mindset is about getting to know a plant and getting better at what you do, a seed steward mindset is to take the plants you work with under your care and protect and cherish them until they are handed over to the next generation of farmers.

At the least you want to leave the varieties you work with at least as good as you received them. But you also have the opportunity to adapt these varieties to where and how you farm. And ultimately you can also create whole new varieties to leave to future generations.

And all of this starts with the following simple step ...

SPEND TIME WITH YOUR PLANTS TO SEE WHO THEY ARE

Becoming a seed steward is about developing a relationship with your plants—and as in any relationship, the way to nurture your relationship with your plants is to spend time with them. (This is probably the most important part of becoming a steward.)

Spend time with your plants—watch them, touch them, taste them, smell them. Just be around them. Do this as part of your work as a farmer, and do this because you simply love their company.

As a seed farmer you can influence how a variety evolves by what plants you encourage to go to seed. So you want to be careful in the plants you choose. The way you learn what plants you want to leave in your seed population and what plants you want to harvest for the market is by spending more time with your plants.

There are so many opportunities to spend quality time with your seed crops.

1. While you plant your crops

When you put a seed in the ground or a seedling tray, you have your first chance to see whether all the seeds look the same or whether some are different from others. It might be harder to tell between different arugula seeds, but the bigger the seed the more easily you'll notice differences.

When those first sprouts push out of the ground you have the first chance to see how vigorous different seedlings are, whether their leaves grow the way you expect, and whether their color and shape is uniform or not.

For those crops that you start in the nursery, you will have to handle each plant one by one to put it into the ground. This gives you another chance to see and choose plants that are sturdier than others and that have more vigor.

Sometimes the differences you see in your young plants might just be from irregular watering, or a bit more fertility here or there, but then again it might not be.

2. While you work in the field

As you walk down the row with your hoe weeding between plants, you have another chance to look at each plant one by one and to also scan the whole population to get a feel for what it's like. As you prune or trellis certain crops, you get your hands on all the plants and get a physical feel for these crops. Even if you're not directly working on a crop, such as when you're irrigating, or driving by with a wagonload of straw bales, you have other opportunities to see your crops.

When you head out to the field for any task, make sure to have flagging tape or flags on you, or on your tractor, or stashed somewhere at the end of some of your field blocks. This way you always have a way to highlight the individual plants that really impress you.

3. While you stroll through your field

Make it a point to take weekly field walks to look at your seed crops where your aim is to look at them and reflect on impressions you had during the week. This can be while you're also making a list for next week's harvest or highlighting weeding priorities. But it can also be a great final thing of the day when all your duties are over—just go and walk the field. Don't forget your flagging tape!

And it's more than just looking. Stop and taste different plants.

4. While you harvest

While you harvest, you are touching each plant once again. Slow down a bit when you're working with the crops that you're growing for seed. Pay attention to what makes for a good bunch or a nice harvest. Look at the carrot stems and how they hold on to a plant as you wrap an elastic around them, or how your peppers are spread out on the plant rather than being all clustered in the center. Pay attention to what you like and then look for other plants that are like that.

Harvest is usually the last moment you have to leave a plant in the field to go to seed. At this point you get to see all your flagged crops again and decide what to leave for seed. And you have one last chance to flag more crops.

For root crops though, this is the first time you see the part people want to eat. Your favorite roots can be put in a separate bin to go for seed next year.

The fruit that comes from any flagged plants can go into separate bins or containers so they don't go to the packhouse to get prepped for market. You can also write notes on your fruits with a sharpie so you don't accidentally mix them up.

Fig. 9.1: *Some peppers I labeled so that I would remember what plant they came from and to make sure they didn't get eaten accidentally.*

5. While you wash and pack your crops

Once your crop is back at the pack house, you can still reflect on what makes a nice lettuce head, or the perfect bunch of beets. And for many crops, while it's too late for those individuals to go to seed, you can carry this knowledge back to the field the next time you are choosing plants.

But for any fruiting vegetables, if you come upon an individual you love at the wash station, you can put that tomato or pepper aside and keep those seeds. You won't know whether it's from a great plant or not but you'll know it produced at least one beautiful fruit.

6. While you stand behind your market stall

You see what stands up to heat and the sun. And you see what clients go right for and what they come back asking about. You can bring all these thoughts back to the field or to your kitchen to influence what you see, touch, and feel.

7. While you pare and trim and cook

Your kitchen is another place to learn what you like and don't. For all your crops, this is knowledge that you can bring back to the field to influence your selections.

For fruiting crops, it gives you another chance to taste and enjoy certain individuals over others. When you clean out the seeds of a squash, put those seeds aside rather than compost or roast them. When the squash comes out of the oven, give it a taste. If it makes your mouth water and your body tingle, go and clean those squash seeds. If it is just average, let them go.

Fig. 9.2: *How well a squash stores in your kitchen can be another selection criterion.*

You can also see which of your fruiting crops store the longest. Keep the seeds from fruit that linger longest on your counter without letting them turn to stinky mush. Add extra points if it tastes great!

Every time you watch, or handle, or taste a plant, you are building a relationship with that crop. The deep knowledge you develop about your plants is what will guide you as you steward your plants and guide them along different paths.

CAREFULLY MAINTAIN THE PLANT CHARACTERISTICS THAT ARE IMPORTANT TO YOU

Starting with a good seed and good varieties is one thing, but your job as a seed steward is to also do what you can to make sure these varieties stay good. This means maintaining a population of plants that are true to type for the qualities and traits that are important to you but that still have the genetic potential to adapt to a changing environment and changing farm needs.

To do this you need to:

1. Isolate seed crops enough to avoid most crossing.
2. Grow enough plants to maintain genetic diversity.
3. Remove plants that move your variety in a different direction.

Isolate Seed Crops Enough to Avoid Most Crossing

Most seed reference books emphasize keeping a really big distance between varieties as critical to maintain a variety. The isolation distances I've been telling you about are a lot closer than in most of the books. I stand by them. Rather than avoiding all cross-pollination, you should reduce most unwanted crossing and keep an eye out for what does cross.

Grow your selfers 50 feet to 100 feet apart to keep crossing to a minimum.

Your intermediate selfers are good with 300 to 600 feet.

Aim for 1,000 to 1,500 feet for your crossers. Try to make sure you're planting varieties of the same species with barriers in between them, such as greenhouses, out-buildings, flowering plants, or tree lines. These barriers will reduce your chance of crossing.

If you're in a commercial seed-growing region, these distances might not be far enough. There is often much more pollen in the air and on insects. In these areas you should expect to use larger isolation distances, especially with crossers.

All that being said, the best way to figure out your isolation distances is to grow your seed crops of the same species with a distance that you think (or hope) is acceptable. Then grow out the plants from the seed you kept and see what they look like.

If you see no crossing, that distance is probably good. If you see crossing, then you need to increase your isolation distances.

If there are low levels of crossing, cull those plants before they go to flower to avoid more crossing. If you're dealing with crops where you only see the crossing at the fruit stage, then you'll have to pick all the current fruit once you see crossing, pull the intruders, and then let the bees carry on with their work. This can be especially frustrating with squash.

Grow Enough Plants to Maintain Genetic Diversity

The biggest change from First Seed Mindset to being a seed steward is how many plants of the same variety you will actually let go to seed.

Most plants have the potential to suffer from inbreeding depression. This happens when plants mate with individuals too closely related to each other. Inbreeding depression results in smaller plants that aren't as resilient or productive. The more susceptible a plant is to inbreeding depression, the more individuals you should grow to have a broader genetic base. Populations that are too small will gradually result in varieties that no longer perform as well.

Generally the ideal population size depends on whether a crop is a selfer or a crosser.

Selfer population size

Selfers don't suffer from inbreeding depression and you don't need as many plants to maintain a good population. You should grow at least 20 plants of a selfer to have a good population where you can maintain any genetic variability. And 60 plants would be even better!

Crosser population size

Crossers suffer more noticeably from inbreeding depression. There is still a range in crossers. You should probably let at least 80 crosser plants go to seed. But you will have much better results if you let 200+ plants go to seed. That is a lot of seed. It might not be easy to justify such big populations on your farm without selling some of this seed.

Cucurbit population size

The cucurbit family is the one big crosser exception in regards to population numbers. A lot of cucurbits have been grown over time in small populations, so they are already kind of pre-inbred. Because of this, you can treat most cucurbits like selfers. So 20 plants would be good, and 60 plants would be better.

Remove Plants That Move Your Variety in a Different Direction

Now that you have a lot of plants that are mostly isolated from other varieties of the same species, the next thing to do is to remove those that are different from what you want your variety to look like.

If everything looks pretty similar one to another, look a little harder! There is always some diversity. You should aim to remove at least five to ten percent of your plants to maintain them the way you want them to be.

If there's a lot of diversity then you should remove much more. This is more likely to happen when you start working with a variety. Over time the plant population should be increasingly the way you want it to be.

If you ever discover you had a significant amount of crossing in a previous generation, you can select more aggressively to clean up your variety. But you can also go back one or two generations and get some of your previous seed to grow out from before the time you suspect the crossing happened.

Your final number of harvested seed plants should be in the population sizes I mentioned above to avoid inbreeding depression. The more plants you grow out, the more plants you can remove from your population. As a market gardener, one way to remove plants is to harvest them and bring them to market! So grow a lot of plants and sell off your culls to fund better genetics on your farm.

IMPROVE AND ADAPT YOUR CROPS TO YOUR FARM AND BIOREGION

Maintaining a variety so it performs the way you want is definitely a good thing. But you can go further in your market garden. When you choose what individuals you let go to seed, you also have the opportunity to improve your varieties so they do better on your farm. You can adapt them to your bioregion so that they are earlier to mature or more resistant to the heat or to the cold. You can also improve their flavor and storage life. Really anything that you value in a variety can be selected for.

Here are a few ways that you can improve and adapt a variety:

Grow a lot of plants

The first thing to do when you want to improve a variety is to grow a lot of plants. This lets you see a lot of diversity.

What is a lot of plants? As a starting point, aim to grow four to five times the population size amounts I listed above, as that will give you a lot more opportunity to get rid of plants. That means 100 to 300 plants of a selfer, and 500 to 1,000 plants of a crosser.

If you grow 1,000 onions and you choose the best 100 to 200 onions for seed, you're keeping 10–20 percent of the crop. This will definitely nudge your variety in the right direction.

Now if you grow 10,000 onions, 100–200 onions is the top one to two percent of your crop. That is an even bigger nudge for the onion to go in the direction you want.

In both of these cases, you're going to wind up with a lot of seed! What do you do with that seed? Chapter 11 will give you some ideas.

Select what you like in your nursery

From the moment you plant seeds in the nursery, you can be looking at germination and seedling vigor, the color of the cotyledons, and then the first true leaves. Each of these moments is a chance to influence a population of plants.

When I start a crop in the nursery that I know I will be selecting for seed, I like to seed three seeds per cell. Most of the seeds will germinate and I'll thin down to one plant per cell. I'll get rid of the plants that are too small or wispy or that don't seem to have oomph. Getting rid of two seeds out three means I've already selected for the best 33 percent of those seeds!

Some of that nursery variability is related to nursery management and uneven watering and fluctuating temperatures, but some of it is genetic.

I also grow at least 30 percent more seedlings in the nursery than I need in the field. If I really want to select a variety I might grow 50 percent more than I need. I only plant out the most vigorous and hardy plants to the field that look like what I want them to look like. This is another selection event that nudges my population as I go along.

Embrace adversity and keep what survives

In addition to all the selecting you can do for your plants, the world also wants to help you out. It will send storms and droughts your way in addition to flea beetles and diseases.

Whenever you have an event that knocks back or destroys a number of your plants, pay attention to those that survived. They might just be lucky but they may also be hardier than the plants that didn't make it. Keep the seed from those plants! They will make your population hardier.

Just choose what you like

One of my first experiences with how much you could change a variety was with some Purple Top White Globe Turnips. One winter we had a bunch of turnips left over in the cold room at the end of the winter. I planted those out to see what turnip flowers look like—it turns out they look just like all the other *Brassica rapa* flowers. I collected the seed. That fall I sowed one row from our seed and one row from leftover seed from the previous year. At harvest, Emily pointed out to me how much better the turnips were from our seed—they were rounder, the purple tops were more vivid, and the leaves had much less disease damage. After one generation we had a better turnip and we hadn't even planned for it!

What we had done though was choose good roots for storage to sell and eat through the fall and winter. Any roots that weren't any good did not make it to this next generation. So simply harvesting what we liked made for a better variety.

Can all varieties be improved?

There are some pretty solid varieties out there that folks have been working on for a long time. Some of them might never get better than what you start with. But most varieties can be improved at least a bit.

Varieties that have a lot of diversity are great selection candidates. Going back to those Purple Top White Globe Turnips we grew out—I later learned that most of the seeds for these turnips are actually used to grow fodder for livestock, and that market farmers are just a secondary client for this seed. I bet that there is not a lot of selection for those turnips going on, which is why we had such success with one generation.

If you're growing an uncommon open-pollinated variety that is hard to find in most seed catalogs, there's a good chance that there's not a lot of work that's been done on this crop for your bioregion. These are great candidates to try to improve and adapt to your farm.

CREATE COMPLETELY NEW VARIETIES

So far you've been culling out any crossed-up plants that you might see appearing in your populations. Now's the time to take a second look at those plants. They are where new varieties come from.

If there is something crossed-up that you like, you should keep the seed from that plant. Then grow that crossed seed in a new area. You'll probably see a lot of diversity in the next generation. Keep what you like the best to go to seed and remove what you don't like by harvesting it for market or your veggie baskets. This is essentially the advice I've given you to maintain and improve your varieties but this time, you're doing this with something a lot less uniform. Over five or six generations your variety will start to show less diversity and stabilize around the traits you're selecting.

Here are some of Tourne-Sol's breeding projects.

Fig. 9.3: *Six of the different types of peppers in the F3 generation. I was selecting for the rounder peppers in the upper right.*

Carrot Bomb Peppers

This began in 2006 as an accidental breeding project. I had grown some Bulgarian Carrot Peppers. These are about three inches long and almost an

Fig. 9.4: *After five generations, we had Carrot Bomb Peppers.*

inch wide. They look like a bright orange pointy cayenne. All the plants looked the same except for one. On one plant, all the fruit were squatter and stockier but still a bright orange. It didn't look like any other pepper in the garden. I tasted it—it was pretty spicy. This was probably an F1 cross that happened accidentally the year before. So I cut the fruit open and kept the seeds.

The next year, I grew a bunch of plants. No two of them had quite the same shaped fruits. This was the F2 generation. I kept the seeds from the small round peppers and grew them out again. As you can see in the picture the F3 still had quite a bit of diversity. And I selected the seeds that way every year.

After five generations of selection, the plants in the population were only producing round orange peppers. I started calling them Carrot Bomb Peppers. They are not quite as hot as the Bulgarian Carrot Peppers. I'm guessing 25,000 to 30,000 Scovilles. These Carrot Bombs also happen to be the earliest peppers we have at Tourne-Sol. This happened by simply choosing the heaviest yielding earliest maturing plants.

Rainbow Tatsoi

This is another accidental breeding story. It started in 2007, when I harvested some Yukina Savoy seed from plants that went to flower a little too close to some Pink Petiole Mix that was in flower at the same time. (Yukina Savoy is like a bigger tatsoi with wrinkled leaves. Pink Petiole Mix is a population of leafy greens with pink and purple stems from Wild Garden Seeds.) When I grew out the Yukina Savoy seed, 10–20 percent of the offspring were obviously crossed up.

The off-types had a Yukina Savoy leaf shape but with a lighter green color and pink stems. This contamination meant we couldn't sell the Yukina Savoy seed as Yukina Savoy but it was also the beginning of a project. I kept seed from all the plants that I could see had been crossed up and I began a breeding population.

For the next couple of years I watched what happened as I kept growing this mix. (I also added a bit more diversity by letting this population cross-pollinate with some tatsoi, Vitamin Greens, and our own Winter Greens Mix.)

I started to notice some plants with round purple tatsoi leaves and pink-purple stems. I decided this was the direction I wanted to take with this mix. I started to dig up my favourite plants and replant them in a separate seed garden to make sure they only pollinated with each other.

We released this variety in our online store when we reached pretty uniform leaf shape, purple outer leaves, and mostly pink stems. There were still green hearts showing up in some of the plants but this made a stunning rainbow display of tatsoi goodness!

One thing I didn't expect happened when we grew this tatsoi in our winter greenhouse when the outside weather was -4°F (-20°C) and inside just above freezing.

Fig. 9.5: *This is one of the early generations after Tokyo Bekana crossed with Pink Petiole Mix.*

Fig. 9.6: (Above left:) *Here is some of the diversity after six years of selection. I kept letting the purple plants go to seed and cut the green plants for veggie baskets.*

Fig. 9.7: (Above right) *Rainbow Tatsoi.*

Fig. 9.8: (Left:) *Not so rainbow in the middle of the winter.*

The leaves came out a deep purple with pink stems. Had I seen this earlier I might have instead called it Purple Winter Tatsoi!

Breeding Ruby-Sol From an F1 Tomato

Two hybrid tomato varieties were the foundation of our Tourne-Sol tomato production. They were both red tomatoes. They tasted great. They produced heavy yields. They were just the right size for our basket members to use in a single session. The skin was just tough enough to travel to market but not so tough that you would notice.

Since the seeds were hybrids, we didn't keep our own seeds and we would purchase them from a seed company. One year we discovered one of the two varieties we relied on had been sold out and the other had a crop failure. Neither variety was available.

We did try other varieties that year and ultimately found some other F1 tomatoes we have adopted as standards for our farm. But with the experience of how precarious our tomato seed supply was, my co-farmer Fredéric Thériault (Fred) launched a project to de-hybridize some of our favourite tomato varieties and create our own favourite reliable varieties with good taste, size, yield, adaptability to caterpillar tunnel production, and earliness

Fred started with our market tomatoes growing in caterpillar tunnels. He flagged the plants with the earliest fruit and those that had the best disease resistance. He harvested his favorites and led the Tourne-Sol team through a taste test. He kept the seeds from the tastiest of his favorite tomatoes and we grew them out the following year. Then Fred repeated the process again. Each generation there were more tasty early tomatoes in the mix.

Fig. 9.9: *Ruby-Sol tomato.* Credit: Sophie Descôteaux

Ruby-Sol is one of Fred's favorite new varieties to emerge from this project. It is an indeterminate early red tomato. It produces saladette size fruit that each have a little nubbin on the bottom. The plants have a distinctive wispy curl to the leaves. Ruby-Sol was selected to grow in caterpillar tunnels or unheated greenhouses, but it has done well in the field too!

We still do grow hybrid red tomatoes but we're growing more and more Ruby-Sol as part of our production.

Black Amethyst Radishes

This last breeding story is an intentional breeding project.

In the fall of 2006, I went to Maine to see Frank Morton speak at the MOFGA (Maine Organic Farming and Gardening Association) Farmer to Farmer Conference. I came home from the conference inspired to plan out a breeding project. At the top my list was to create a black winter radish with red flesh.

Next fall, I selected some Black Radishes and some Watermelon Radishes to be the parents of my cross. I overwintered these radishes in the cold room. In the spring, I planted six Black Radish besides six Watermelon Radish. They quickly bolted and the bees went to work. In August, I harvested the seed. I kept the Black Radish seed separate from the Watermelon Radish seed.

Fig. 9.10: *The F1 plants from my Watermelon Radish x Black Radish cross.*

When I planted that seed, most of the roots that grew from the Black Radish seed were Black Radish. Most of the roots from the Watermelon Radish were Watermelon Radish. Some of each planting was different. I had a dozen clearly crossed-up Watermelon x Black Radish F1. They all looked fairly similar. Purple skin, purple flesh, and a bit of black on the outside. This was not the radish I dreamed about.

Next summer I replanted the F1 radishes and harvested their F2 seed. The roots that grew out of that F2 generation was a bonanza of radish diversity with different skin color,

Fig. 9.11: *Some of the F2 bonanza of radish diversity.*

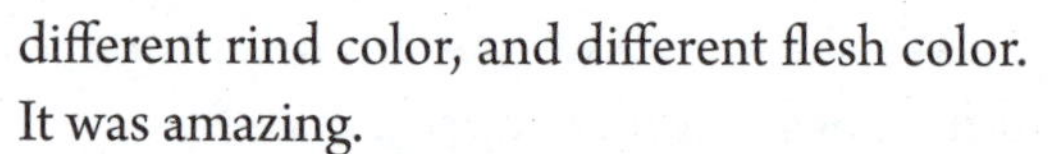
different rind color, and different flesh color. It was amazing.

Among all these radishes were some roots with black skin and red flesh. This was what I wanted. From this point onward it was just a matter of selecting the radishes closest to what I wanted year after year. I've been selecting these for years and I have an almost stable Black Crimson Radish with black skin and red flesh.

But even intentional projects can produce unexpected results. As I was selecting my radishes, I started to find some black radishes with purple flesh! So I also have an almost stable Black Amethyst Radish!

Fig. 9.12: *We cut a slit into every black radish to make sure the flesh is indeed red before planting it out.*

Fig. 9.13: *Black Amethyst Radish.*

Chapter 10

Crop Planning for Seed Farmers

As you get more experienced as a seed farmer, you're going to start growing more crops and varieties for seed and you'll be hoping to harvest larger and larger amounts of seed. To do this effectively you're going to need a good crop plan to make sure you can keep everything coordinated.

I love spreadsheets for crop planning. For this chapter we're going to use my Seed Farmer Crop Plan spreadsheets to plan your seed crops for the coming growing season. You can get the spreadsheets and watch some videos on how to use them at sheets.seedfarmerbook.com. (I actually have a parallel life where I run the Farmer Spreadsheet Academy. Head to www.farmerspreadsheetacademy.com if you want to jump down that rabbit hole.)

This chapter begins with the sales you hope to make from your farm and figuring out how much revenue you need to generate per area to hit that target.

Next is the actual crop planning. You'll go through five steps where you start with how much seed you want to harvest and then work backwards to figure out how many plants you need to grow to do that and when and where you should plant them on your farm. If you already do much crop planning for your vegetable or flower crops, then you'll recognize a lot of the thinking behind this chapter.

There is a section in the Seed Farmer Crop Plan spreadsheets for each of these steps. To plan each of your crops, you can fill out each section of the spreadsheet from left to right. The decisions you make in one section will guide your choices for the next section.

I've illustrated each step with a few examples of how I would go about each of these steps to plan two annual crops: brassica greens and tomatoes. After we've gone through all the sections, there will be one more example to illustrate how this process changes when you're planning for a biennial that you dig up and then replant in another place. For this, I will plan out some beets.

Once we've looked at how crop planning looks on an annual basis, we'll jump into how to plan your crop rotation over multiple years. Then we'll wrap up by

Weight Units and Distance Units in Your Spreadsheets

One secret to not getting confused by your spreadsheets is to be consistent in the measurement units you use. There are two types of measurement units to consider: distance and weight.

Distance units

You can use feet or meters to measure distance, but you should not switch between the two in your spreadsheets. There are two sets of spreadsheet templates for you to choose from at sheets.seedfarmerbook.com: one in feet and one in meters.

Weight units

Measuring weights is not simply a matter of choosing between pounds and kilograms, you actually need to choose between pounds, ounces, kilograms, and grams. These spreadsheets will only work if you use one of the four measurements. Interestingly, your distance and weight units don't have to both be in imperial units or metric units.

The spreadsheet examples in this chapter actually use feet and grams. This is how I think! I like that most market crops are planted on one-foot spacing. I like how much easier math is with grams instead of converting ounces to pounds.

evaluating whether the crops you've planned are able to meet the financial goals you set, and how you can influence your crop profitability.

If you're new to crop planning and this chapter is a bit overwhelming, feel free to take this slowly and only do as much as you wish. Having part of a plan is better than no plan at all!

CALCULATE YOUR SALES PER GROWING SPACE BENCHMARK

If you're planning on earning part of your living from growing seed, you need to know how profitable your seed crops are compared to other crops you could be growing.

To get a good idea of how profitable a crop is, you can calculate a sales per growing area benchmark and then compare your crops to this benchmark. This is not as accurate as a full crop budget where you track all your income and all your expenses associated with that crop, but a sales per growing area benchmark is a lot simpler and quicker to use.

You can use the "Space Benchmarks" tab of the Seed Farmer Crop Plan spreadsheet to calculate the benchmarks for your farm.

How much do you want to sell this year from the field?

This is the gross sales you plan on making from your field operation. This would include vegetable sales, cut flower sales, and bulk seed sales. It does not include any revenue you might make from a retail seed operation.

If you run a retail seed operation, you should consider that you're contracting seed from your own farm. You should only include the bulk value of any seeds you're growing as part of your gross sales for this calculation.

How much growing space do you have?

This is the total area cultivated that is used to produce a crop for sale during the year. Do not include any full-year cover-cropped or fallow areas.

At Tourne-Sol, though we have 12 cultivable acres, we only use eight acres for crop production in any given year. The other four acres are always in cover crop. So we would use eight acres as the amount of growing space for this equation.

What is your bed width from the middle of one path to the middle of the next?

Measure your bed width from the middle of one path to the middle of the next pass. This effectively means that you will be measuring your growing space and one full path.

Calculate your sales per growing space benchmark

Here's the equation to measure your growing space benchmark per acre or hectare:

Gross sales + cultivated acres = $ per acres
Gross sales + cultivated hectares= $ per hectare

If you want to make $100,000 from 2 acres (0.81ha), that would be
$100,000 / 2 acres = $50,000/acre
$100,000 / 0.81ha = $123,500/ha

Thinking in terms of $ per acre or hectare might not be obvious to you. Here's how to convert these benchmarks into $ per bed-foot or bed-meter.

$ per acres × bed width (ft) + 43,560 sq. ft. = $ per bed-foot
$ per hectare × bed width (m) + 10,000 m2 = $ per bed-meter

Here's what $100,000 from 2 acres (0.81ha) looks like if you're growing on 5ft (1.5m) bedwidth.

$50,000/acre × 5ft + 43,560 sq. ft. = $5.74/bed-foot

$123,500/ha × 1.5 m + 10,000 m2 = $18.83/bed-meter

Table 10.1

	Farm A	Farm B
$ per acre	$50,000	$30,000
$ per ha	$123,500	$74,100
$/bed-foot	$5.74	$3.44
$/bed-meter	$18.83	$11.30

It is important to do these calculations for your own farm and not assume someone else's numbers work for you. Here are the results for two farms. Farm A wants to produce $100,000 from two acres (0.81ha). Farm B wants to produce $180,000 from six acres (2.43 ha). You can see they result in different benchmarks

Compare your estimated $/bed-foot or $/bed-meter with your growing space benchmark

In Step 2 below you will estimate your $/bed-foot or $/bed-meter in Column J of the "Crop Plan" tab.

You can compare this estimate with your growing space benchmark. At the end of this chapter we'll explore some options you can use to increase the $/bed-foot or $/bed-meter your seed crops produce.

STEP 1—CHOOSE YOUR CROPS

The white columns of the Seed Farmer Crop Plan spreadsheet are where you record what you will grow. The most important decision to make here is how

Table 10.2

A	B	C
Crop	Variety	Species
Brassica greens	Tokyo Bekana	*Brassica rapa*
Brassica greens	Yukina Savoy	*Brassica rapa*
Brassica greens	Ruby Streaks	*Brassica juncea*
Tomato	Montreal Tasty	*Solanum lycopersicon*
Tomato	San Marzano	*Solanum lycopersicon*
Tomato	Jaune Flammée	*Solanum lycopersicon*
Tomato	Galina	*Solanum lycopersicon*

- Col A: Crop: what crops you are growing.
- Col B: Variety: the specific variety of each crop you're growing.
- Col C: Species: the species of each crop you want to grow.
- There will be a different row for each variety you grow.

many varieties of the same species can you grow in the space you have. You'll need to consider what isolation distances you need to keep cross-pollination within an acceptable range. If you're selling seed on contract this is especially important.

For these crop planning examples I want to grow some brassica greens and tomatoes for seed.

At Tourne-Sol farm, there are two isolation plots that are about 1,200 feet apart. In our experience on our farm this is enough distance to let us grow a crosser of the same species in each block without seeing any crossing. There are also four smaller plots that are about 300 feet apart from each other. With this layout, we can grow up to two crossers of the same species and six selfers in a given year.

Tomatoes are selfers. That means I could grow up to six varieties in these current isolations. I'll choose to grow four varieties this year. I enter four rows of tomatoes into the spreadsheet: Montreal Tasty, San Marzano, Jaune Flammée, and Galina.

I also want to grow three different varieties of brassica greens. I know they are crossers so I look up the species. Two of these varieties are Brassica rapa: Tokyo Bekana and Yukina Savoy; one variety is Brassica Juncea: Ruby Streaks. I can grow the two Brassica rapas in separate plots that are 1,200 feet apart without fear of crossing. I can grow the Brassica juncea right beside one of the Brassica rapas without worry. I enter the three brassica green varieties as three separate rows in the spreadsheet.

STEP 2—PLAN YOUR HARVEST NEEDS

The peachy-orange columns (F, G, H, I, and J) of the Seed Farmer Crop Plan spreadsheet are where you enter how much seed you want to harvest. At this point, you can also estimate the monetary value of each seed crop.

How Much Seed Do You Want to Harvest?

Consider each way that you use seed on your farm and evaluate the seed needs.

Seed for farm use

Go and look at how much seed you use to grow that crop for market. You can find this information in your annual seed order (see Chapter 1) or the seed needs from your crop plan. This is the amount of seed you use in one year.

- Col D and E are used to plan for vegetable or flower successions but not for seed crop planning.
- Col F: Harvest Units: This is how you will measure your harvest. Vegetable harvest might be measured by weight, or bunch, or each individual unit. The final seed harvest is usually measured in pounds, ounces, kilograms, or grams
- Col G: Harvest Needs: Your harvest goal for this crop. Measured in the Col E harvest units.

Table 10.3

A	B	F	G
Crop	Variety	Harvest Units	Harvest Needs
Brassica greens	Tokyo Bekana	g	10,000
Brassica greens	Yukina Savoy	g	10,000
Brassica greens	Ruby Streaks	g	4,000
Tomato	Montreal Tasty	g	500
Tomato	San Marzano	g	500
Tomato	Jaune Flammée	g	500
Tomato	Galina	g	500

As mentioned in Chapter 8, most seeds will last many years. You should plan on growing enough seed for two to three years of farming. Not growing all your varieties out each year means you can grow a different variety in the other years.

If you use 1 lb (454 g) of Tokyo Bekana each year, then plan on growing three times that amount for farm use.

1 lb (454 g) × 3 = 3 lbs (1,362 g)

Seed for packet sales

If you've got past sales records, start with the amount of packs you sell per year. If you generally ran out of seed for that variety, forecast the amount of packs you could have sold with more seed available. Estimate what you might sell over two or three years.

If you sell 300 packs of Tokyo Bekana each year and you put 1/16 oz (1.75 g) of seed per packet, then you need 18.75 oz (525 g) per year.

300 packs × 1/16 oz (1.75 g) per pack = 18.75 oz (525 g)

Over three years, you hope your pack sales will go up by 20% so you need:

3 years × 18.75 oz (525 g) per year × 1.2 = 67.5 oz or 4.2 lbs (1,890 g)

Seed for a seed contract

This one's easy—the amount that your contract specifies! With seed contracts, you don't want to grow more than one year's seed. There is no guarantee that you'll have the market for that seed next year!

If you have a contract for 15 lbs (6,810 g), then use that amount!

Add everything up!

The total of those three amounts are what you put in column G of your spreadsheet.

Table 10.4

Needs	Lbs	Oz	Grams
Farm Use	3	48	1,352
Packet Sales	4.2	67.2	1,890
Seed Contract	15	240	6,810
Total	22.2	355.2	10,052

Estimate the Monetary Value of Your Crop

These columns calculate the value of the bulk seed you are growing. Be careful with this number when you're growing seeds—it only corresponds to actual sales if you sell the seed.

Let's say I sell bulk Tokyo Bekana seed for $140/kg ($63.64/lb) which is also $0.14/g. I entered the price per gram in the spreadsheet because that was the unit I set in Col F.

Multiplying my 10,000 g harvest by $0.14/g means I'm expecting to produce $1,400 of bulk seed.

Multiplying the price per unit of $0.14/g times by my yield estimate of 30g/bedft means that each bedft used to grow Tokya Bekana is expected to generate $4.20/bedft at bulk seed prices.

Table 10.5

A	B	H	I	J
Crop	Variety	Price per unit	Sale forecast	Estimated $/bed-foot
Brassica greens	Tokyo Bekana	$0.14	$1,400.00	$4.20
Brassica greens	Yukina Savoy	$0.14	$1,400.00	$4.20
Brassica greens	Ruby streaks	$0.14	$560.00	$4.20
Tomato	Montreal Tasty	$1.00	$500.00	$12.00
Tomato	San Marzano	$1.00	$500.00	$5.00
Tomato	Jaune Flammée	$1.00	$500.00	$12.00
Tomato	Galina	$1.00	$500.00	$18.00

- Col H: Price per unit: This price needs to be in the same units as Col F.
- Col I: Sale Forecast: This is calculated by Col G (harvest needs) × Col H (price per unit).
- Col J: Estimated $/ bedft ($/bedm): This is calculated by Col H (price per unit) × Col R (yield estimate).

STEP 3—PLAN YOUR FIELD PLANTINGS

The cyan blue columns (K, L, M, N, O, P, Q, W, and X) of the Seed Farmer Crop Plan spreadsheet are where you decide what will happen in the field. This is the step where you will spend the most time crop planning. There are four decisions to make:

1. When you will plant your crops
2. What crop spacing you will use
3. How much of each crop you will plant
4. Where you will plant your crops on your farm.

When to Plant Your Crop?

Many vegetable growers' first attempt at keeping seeds happens with vegetables that go to flower before they can be harvested for market. This is just what happened to me in my arugula story in the Introduction. Sometimes this works out, but you might also find that if a crop goes to flower too late in the season there just won't be enough time for it to mature. You need to plant seed crops at the right time to get a reliable harvest.

The right time to plant seed crops depends first off on how hardy the crop is. You should generally aim to plant as early as you can in that planting window to make sure your crop has enough time to mature its seeds.

Four general planting periods for seed crops

Here are some general guidelines for when to schedule different seed crops. I have added what dates we aim for at Tourne-Sol (USDA zone 4, Canadian Hardiness zone 5). Your actual planting dates will depend on what climate you are in.

- When soil is dry enough to plant and major frosts have passed:
  - Plant hardy annual seed crops.
  - Plant first-year onions that will produce bulbs.
  - Replant second-year biennials that have overwintered in cold storage.
  - *At Tourne-Sol we aim to plant these crops the first week of May.*
- When the last chance of frost is passed and the soil has warmed up:
  - Plant tender annual seed crops.
  - *At Tourne-Sol we aim to plant these crops the first week of June.*

- Mid–summer:
  - Plant first-year biennials so that they are not too big going into the winter.
  - The warmer your climate the wider a range you have for these crops.
  - *At Tourne-Sol we aim to plant these crops in mid-July to early August.*
- End of summer
  - Plan overwintering hardy annuals to produce seed the next year.
  - *At Tourne-Sol we aim to plant these crops in early September.*

For my example crop, I write that the brassica greens are hardy crops in Col L. I plan on planting them on May 6. I write down that tomatoes are a tender crop in Col L. I plan on planting these on June 3.

Table 10.6

A	B	K	L	M	N
Crop	Variety	Days to Maturity	Field Date Target	Field Date Plan	Field Week
Brassica greens	Tokyo Bekana	n/a	Hardy	2024-05-06	19
Brassica greens	Yukina Savoy	n/a	Hardy	2024-05-06	19
Brassica greens	Ruby streaks	n/a	Hardy	2024-05-06	19
Tomato	Montreal Tasty	n/a	Tender	2024-06-03	23
Tomato	San Marzano	n/a	Tender	2024-06-03	23
Tomato	Jaune Flammée	n/a	Tender	2024-06-03	23
Tomato	Galina	n/a	Tender	2024-06-03	23

- Col K is used to plan for vegetable or flower successions but not for seed crop planning.
- Col L: Field Date Target: Use this column to enter the plant hardiness group.
- Col M: Field Date Plan: Choose the date that you want to actually plant this crop.
- Col N: Field Week: This formula calculates what week of the year for the date in Col M. Week 1 is the week of January 1.

Choose Your Crop Spacing

This is where you define how you will grow your crops in the field. When choosing your crop spacing you should consider the factors from Chapter 2, especially ventilation and weed control.

I will plant brassica greens three rows per bed to make sure they have enough ventilation as they grow and to make it easy to weed them through the season. I will direct seed both the Tokyo Bekana and Ruby Streaks aiming for 15 seeds per bed-foot (49 seeds per m) for these crops. I will transplant the Tokyo Bekana with 1 ft (0.3m) spacing between each plant.

Table 10.7

A	B	O	P	Q
Crop	Variety	Rows/ Bed	Transplant In-Row Spacing (feet)	Direct Seed (seeds/ft)
Brassica greens	Tokyo Bekana	3	0	15
Brassica greens	Yukina Savoy	3	1	0
Brassica greens	Ruby streaks	3	0	15
Tomato	Montreal Tasty	1	1.5	0
Tomato	San Marzano	1	1.5	0
Tomato	Jaune Flammée	1	1.5	0
Tomato	Galina	1	1.5	0

- Col O: Rows /Bed: The number of rows for this crop you plant per bed.
- Col P: Transplant In-Row Spacing (feet or meters): This is the space between crops that are transplanted.
- Col Q: Direct Seed (seeds/ft or seeds/m): This is the number of seeds in a given amount of row. This should be based on the actual seeder you use.
- A crop can only have a value greater than 0 in one of Col P or Col Q. It cannot be both direct-seeded and transplanted at the same time.

How Much of Each Crop Will You Plant?

This is a critical question for vegetable and flower growers. Vegetables and flowers are very perishable so if you grow more crop than your market can handle, you're essentially growing compost. You should err a bit on the conservative side when you're planning how much of a vegetable or flower crop to grow.

Seeds are a lot less perishable than vegetables or flowers. This means that you could grow an overabundance of seeds and still be able to use, share, or sell the harvest over a number of years. There are a few seed crops like parsnips that don't store for very long, but they're the exception rather than the rule.

That being said, it's still a good idea to figure out mathematically how many seeds you should grow. You can do that with this equation:

> Harvest needs × field SF ÷ yield = bed-feet or bed-meter to plant.
>
> *I want to grow 10,000 g of Tokyo Bekana seeds. I estimate that I can harvest 30g/bed-foot (entered in Col S) and I want to plan around a 1.3 safety factor (entered in Col S.) The spreadsheet does the following math:*
>
> *10,000g × 1.3 Safety Factor ÷ 30g per bed-foot = 433 bed-feet*
>
> *I round off 433 bed-feet to 450 bed-feet for my crop plan and enter that in Col U*
>
> *Here's the math in other units:*
>
> *10,000 g × 1.3 Safety Factor ÷ 98 g per bed-meter = 132 bed-meter*
> *357 oz × 1.3 Safety Factor ÷1.07 oz per bed-foot = 433 bed-feet*

Yields

I have provided yields in the Part 2 crop profiles and in the chart in Appendix 4. These yields are meant to be a starting point for your planning. Estimating seed yield can be difficult. Seed yields vary greatly from climate to climate, and even more importantly from variety to variety. The best yield information comes from keeping track of how much seed you harvest from a certain area. Averaging this out over a couple of years will provide the best yields for your own planning.

Safety factors (SF)

You know that not every plant is going to make it. Weather, creatures, weeds, and disease will all take some of your plants and reduce your yield. Your field safety factor helps offset these vagaries. Your field safety factor also offsets the fact that your yield estimate is just an estimate.

I generally recommend planning around a 30 percent field safety factor. This corresponds to 1.3 in your spreadsheets.

Be aware that if everything works out well in the growing season, you might wind up with 30 percent more harvest than you expected. When your seed is for on-farm use or seed pack sales, this extra harvest shouldn't be a problem. If you're growing large volumes of a seed variety on contract, a 30 percent surplus might be a lot. If you're on contract, be sure to discuss with your buyer what happens with extra seed. How much extra will they buy from you and at what price? Can you sell any extra seed to other seed companies?

How many plants are you growing?

Once you've figured out how many bed-feet, you can calculate how many plants you will be growing of that crop.

For a transplanted crop the equation is

Bed-feet × rows ÷ row spacing = number of plants

Bed-meter × rows ÷ row spacing = number of plants

To grow 450 bed-feet of transplanted Yukina Savoy: 450 bed-feet × 3 rows/bed ÷ 1 ft/plant/row = 1,350 plants.

For a direct-seeded crops, the equation actually gives you the number of seeds sown rather than plants. But you can use this as your number of plants:

Bed-feet × rows × seeds/ft = number of seeds

Bed-meters × rows × seeds/m = number of seeds

To grow 450 bed-feet of direct-seeded Tokyo Bekana: 450 bed-feet × 3 rows/bed × 15 seeds/ft = 20,250 seeds.

Table 10.8

A	B	R	S	T	U	V
Crop	Variety	Yield Estimate (unit/bed-foot)	Field Safety Factor	Bed-feet Needs	Bed-foot Plan	# of plants or seeds
Brassica greens	Tokyo Bekana	30.0	1.3	433.3	450	20,250
Brassica greens	Yukina Savoy	30.0	1.3	433.3	450	1,350
Brassica greens	Ruby Streaks	30.0	1.3	173.3	150	6,750
Tomato	Montreal Tasty	12.0	1.3	54.2	50	33
Tomato	San Marzano	5.0	1.3	130.0	125	83
Tomato	Jaune Flammée	12.0	1.3	54.2	50	33
Tomato	Galina	18.0	1.3	36.1	50	33

- Col R: Yield Estimate (unit/bed-foot or unit/bed-meter).
- Col S: Field Safety Factor: A safety factor to account for all the plants you might lose during the growing season.
- Col T: Bed-feet Needs or Bed-meter Needs: a formula that calculates how many bedft or bedmeter you need to meet your harvest needs.
- Col U: Bed-foot Plan or Bed-meter Plan: Choose the number of bed-feet or bed-meters you will actually grow.
- Col V: Number of plants or seeds. A formula that calculates how many plants to transplant or seeds to direct seed for the bed length you chose in Col U.

Take a moment to look at the number of plants and think back to Chapter 9 where we talked about population size. How does the number of plants you're growing compare with recommended population sizes? Is this enough plants for your stewardship goals?

Where Will You Plant?

These columns let you choose where in your field you will actually plant these crops.

At Tourne-Sol we label our blocks with 3-digit codes, so 101, 108, 116, and 201 are different blocks in our fields.

Table 10.9

A	B	W	X
Crop	Variety	Block	Bed #
Brassica greens	Tokyo Bekana	116	1 + 2
Brassica greens	Yukina Savoy	201	1 + 2
Brassica greens	Ruby Streaks	116	2
Tomato	Montreal Tasty	116	4
Tomato	San Marzano	201	4
Tomato	Jaune Flammée	101	1
Tomato	Galina	108	1

- Col W: Block
- Col X: Bed Number

STEP 4—PLAN YOUR NURSERY SEEDING

The frog-green columns (Y, Z, AA, and AB) of the Seed Farmer Crop Plan spreadsheet are where you plan out what you need to do in your nursery. Though there are many columns in this section, they are much simpler decisions to make than your Step 3 decisions. Mainly you want to make sure you know when to start seeds in your nursery, how much to seed, and how much to pot up.

When Should You Start Plants in the Nursery?

There are two dates you can plan for in the nursery: when you will seed a crop into a cell tray, and when you will pot up a seedling into a larger cell or pot.

> *I will grow tomatoes for seven weeks in the nursery. In Col N, I can see I should plant seedlings into the field on week 23. That means that the tomatoes need to be seeded on week 16. When these tomato seeds have germinated and produce their first true leaves, I would like to transfer these plants into larger cells to give them more space to grow. I estimate they will be big enough to pot up three weeks after they are seeded. I expect the tomatoes will be potted up*

Table 10.10

A	B	Y	Z	AA	AB
Crop	Variety	Weeks In Nursery	Seeding Week (Date)	Weeks Before Pot-Up	Pot-Up Week (Date)
Brassica greens	Tokyo Bekana		DS Crop		No Pot Up
Brassica greens	Yukina Savoy	4	15		No Pot Up
Brassica greens	Ruby Streaks		DS Crop		No Pot Up
Tomato	Montreal Tasty	7	16	3	19
Tomato	San Marzano	7	16	3	19
Tomato	Jaune Flammée	7	16	3	19
Tomato	Galina	7	16	3	19

- Col Y: Weeks In Nursery: The number of weeks from the moment you seed a crop to the moment it will be planted in the field.
- Col Z: Seeding Week (Date): This is calculated by Col N (field week) minus Col Y (weeks in nursery)
- Col AA: Weeks Before Pot-Up: The expected number of weeks from the moment you seed a crop to the moment it will be potted up.
- Col AB: Pot-Up Week (Date): This is calculated by Col Z (seeding week) + Col AA (weeks before pot up)

on week 19. If they are growing quicker than planned, I will pot them up a bit earlier. If they are growing slower, I will pot them up a bit later.

For Yukina Savoy, I plan on four weeks in the nursery. These seedlings are planned for the field on week 19. That means they need to be seeded on week 15. These seedlings will not be potted up, they will stay in the same cells the whole time they are in the nursery.

How Many Trays Should You Pot Up?

Here is the equation to calculate how many plants you need to pot up:

# of plants for the field × pot-up SF ÷ pot-up tray size = # of trays to pot up

A ten percent safety factor (1.1 in Col AC) is adequate when you're potting up seedlings. At this point your seeds have already germinated and are growing in a protected nursery. You will generally lose very few plants at this time.

I want 83 San Marzano tomato plants for the field. I will pot up seedlings into three-inch pots that fit 18 pots in a tray. I calculate 83 plants × 1.1 SF ÷ 18 plants/ tray = 5.1 trays. I round this number down to 5 trays in Col AF because it is easier to manage full trays in the nursery.

Table 10.11

A	B	AC	AD	AE	AF
Crop	Variety	Pot-Up Safety Factor	Pot-Up Tray Size	Pot-Up Tray Needs	Pot-Up Tray Plan
Brassica greens	Tokyo Bekana			0.0	
Brassica greens	Yukina Savoy			0.0	
Brassica greens	Ruby Streaks			0.0	
Tomato	Montreal Tasty	1.1	18	2.0	2
Tomato	San Marzano	1.1	18	5.1	5
Tomato	Jaune Flammée	1.1	18	2.0	2
Tomato	Galina	1.1	18	2.0	2

- Col AC: Pot-Up Safety Factor: A safety factor to account for all the plants you might lose after they are potted up.
- Col AD: Pot-Up Tray Size: The number of cells per tray of the tray you are potting up in.
- Col AE: Pot-Up Tray Needs: A formula that calculates how many trays you will seed to get the number of plants in Col V.
- Col AF: Pot-Up Tray Plan: Choose the number of trays you actually want to pot up.

How Many Trays Should You Seed?

There are two equations to calculate how many trays you should seed in the nursery.

I recommend a 30 percent safety factor (1.3 in Col AG) when you're seeding in the nursery to make sure enough seed germinate and become strong seedlings.

Tray needs for crops that are not potted up

of plants for the field × seeding SF ÷ seeding tray size = # of trays to seed

I want to plant 1,350 Yukina Savoy plants into the field and want to calculate how many 128-cell trays I should start to have that number of plants: 1,350 plants × 1.3 seeding SF ÷ 128 cells/tray = 13.7 trays. I round this number to 14 trays in Col AK.

Tray needs for crops that are potted up

of trays you plan to pot up × pot-up tray size × seeding SF ÷ seeding tray size = # of trays to seed

I want to pot up five trays of San Marzano tomato plants. Each potted-up tray will contain 18 pots. I calculate how many 128-cell trays to start to have enough plants to pot up: 5 trays × 18 pots/tray × 1.3 seeding SF ÷ 128 cells/tray = 0.9 trays. I round this number to 1 tray in Col AK.

Table 10.12

A	B	AG	AH	AI	AJ	AK
Crop	Variety	Seeding Safety Factor	Seeding Tray Size	Seeds/ Cell	Seeding Tray Needs	Seeding Tray Plan
Brassica greens	Tokyo Bekana				0.0	
Brassica greens	Yukina Savoy	1.3	128	2	13.7	14.0
Brassica greens	Ruby Streaks				0.0	
Tomato	Montreal Tasty	1.3	128	1	0.4	0.5
Tomato	San Marzano	1.3	128	1	0.9	1.0
Tomato	Jaune Flammée	1.3	128	1	0.4	0.5
Tomato	Galina	1.3	128	1	0.4	0.5

- Col AG: Seeding Safety Factor: A safety factor to account for all the plants you might lose before they are transplanted to the field or potted up.
- Col AH: Seeding Tray Size: The number of cells per tray of the tray you are seeding in to. If you use open flats, this is the number of seeds per tray.
- Col AI: Seeds /Cell: How many seeds you sow per cell to offset dormancy or low germination and guarantee you get at least one seedling. In most cases you will thin your plants down one seed per cell if multiple seeds germinate.
- Col AJ: Seeding Tray Needs: There are two different formulas nested in this column: one for crops that are potted up, and one if they are not potted up. A formula calculates how many trays you will seed to get the number of plants in Col V.
- Col AK: Seeding Tray Plan: Choose the number of trays you actually want to grow based on the space and resources you have.

STEP 5—CALCULATE YOUR SEED NEEDS

The lilac-purple columns (AL and AM) of the Seed Farmer Crop Plan spreadsheet are where you calculate how many seeds you need to have on hand to be able to follow your crop plan.

There are two different formulas to calculate how many seeds you need, depending on whether you are direct seeding your crop or transplanting it.

In both cases, I recommend using a 30 percent SF (use 1.3 in Col AL) to make sure you have enough seeds on hand when it's time to sow.

Seed needs for crops that are direct seeded

of plants × seed order SF = seed needs

I plan on seeding 20,250 plants (Col V) of Tokyo Bekana. I will need: 20,250 plants × 1.3 SF = 26,325 seeds.

Seed needs for crops that are transplanted

of trays you plan to seed × seeding tray size × seeds/cell × seed order SF = seed needs

I plan on seeding 14 trays of 128 cells of Yukina Savoy with 2 seeds per cell. I will need: 14 trays × 128 cells/tray × 2 seeds/cell × 1.3 SF = 4,659 seeds.

- Col AL: Seed Order Safety Factor: One more safety factor.
- Col AM: Seed Needs: The total number of seeds you need to grow the plants that will hit your seed harvest needs.

Table 10.13

A	B	AL	AM
Crop	Variety	Seed Order Safety Factor	Seed Needs
Brassica greens	Tokyo Bekana	1.3	26,325
Brassica greens	Yukina Savoy	1.3	4,659
Brassica greens	Ruby Streaks	1.3	8,775
Tomato	Montreal Tasty	1.3	83
Tomato	San Marzano	1.3	166
Tomato	Jaune Flammée	1.3	83
Tomato	Galina	1.3	83

CROP PLANNING FOR REPLANTED BIENNIALS

The crop planning steps described above work smoothly when you're planning crops that stay in the same spot from the moment they are planted to the moment they are harvested. If you plan on digging up your biennials and replanting them somewhere else you'll need to consider your seed crop as two separate crops in your crop plan. A year-one crop and a year-two crop. For planning purposes you should start with the final year-two crop and then work backwards for the year-one crop.

> *I want to grow some Touchstone gold beet seed. I will grow these over two years. In year one I will seed the beets in mid-July and harvest the roots in early October to overwinter in cold storage. In year two, I will plant the overwintered roots into the field in early May and harvest the seeds in August.*

Plan your year-two biennial

Start with your seed harvest needs and figure out how many plants you will need to replant. In this case you can consider Col AM (Seed Needs) as the number of plants you will need to harvest in year one.

> *I want to grow 10,000 g of Touchstone gold beet. First, I calculate my bed-feet needs: 10,000 g × 1.3 Safety Factor ÷ 75 g per bed-foot = 173 bed-feet. I round this up to plant 200 bedft of overwintered roots in year two.*

I calculate how many plants I will need: 200 bed-feet × 1 row/bed ÷ 1 foot/plant/row = 200 plants.

And then, how many roots I would need to harvest in the previous fall: 200 roots ×1.3 SF = 260 roots. I round this up to harvest 300 roots in year one.

Plan your year-one biennial

You start with the plants you want to harvest in year one and calculate how many bed-feet to plant and how many seeds you need to start with.

I calculate the bed-feet needed to produce 300 roots. I estimate I will be able to harvest ten nice roots per bed-foot: 300 roots × 1.3 SF ÷ 10 roots/bed-foot = 39 bed-feet. I round this up to 50 bed-feet.

I then calculate how many seeds I will need to sow: 50 bed-feet × 3 rows/bed × 20 seeds/ft/row = 3,000 seeds.

I multiply my seed needs by a 1.3 safety factor to make sure I have enough seeds on hand: 3000 seeds × 1.3 SF = 3,900 seeds.

Table 10.14

A	F	G	R	S	T	U	V	AL	AM
Crop	Harvest Units	Harvest Needs	Yield Estimate (unit/ bed-foot)	Field Safety Factor	Bed-feet Needs	Bed-foot Plan	# of plants or seeds Factor	Seed Order Safety	Seed Needs
Beets Year 2	g	10,000	75.0	1.3	173.3	200	200	1.3	260
Beets Year 1	roots	300	10.0	1.3	39.0	50	3,000	1.3	3,900

CROP ROTATION FOR SEED FARMERS

Up to now, we've only considered planning crops on an annual cycle. On a multi-year scale, you want to avoid planting crops from the same botanical family in the same spot year after year. Crops that are closely related tend to be susceptible to the same diseases. If you grow the same crop families in the same spot year after year, these diseases will accumulate in your soil. To break these disease cycles you need a crop rotation plan.

If you're only growing a few seed crops as part of a larger market garden, you can simply integrate these seed crops into your market garden crop rotation.

If you're growing many seed crops, you'll probably find it is easier to create a rotation around your seed crops.

Here is a simple seed crop rotation that is easy to adapt to many situations. There are two principles guiding this rotation: seed crops are grouped by plant hardiness, and seed crops alternate with cover crops.

Group Seed Crops by Plant Hardiness

The rotation is built around grouping crops in plant hardiness groups: frost-hardy crops and tender crops. Generally tender crops and hardy crops are in different families, so alternating these two groups will break up disease cycles.

Frost-hardy crop blocks

These blocks will be planted once there is no more chance of heavy frosts in the spring. These crops can usually handle light frosts. Crops in this group include annual brassicas, peas, lettuce, and many flowers. It also includes replanted year-two biennials.

Tender-crop blocks

This block will be planted after all chance of frost has passed and your soil warms up. It includes tomatoes, peppers, eggplants, squash, cucumbers, melons, beans, corn, and many flowers.

Crop families to be careful with

There are a few families that contain both hardy and tender crops: primarily the *Asteraceae* family (lettuce and a lot of flowers), and the *Fabaceae* family. Be extra careful with *Asteraceas* since they are very susceptible to diseases. Sclerotinia, downy mildew, and a few other diseases can hit multiple plants in this family pretty hard. These diseases can linger in soil for a while, building up and creating problems. Fabaceae crops don't have as many challenges.

Alternate Seed Crop and Cover Crops

After each year of seed crops, there is a cover crop. During your cover crop year, you can flush out any shattered seeds that could become weeds. During the cover crop year, the shattered seed germinates and starts to grow. When you mow the cover crops, you also keep all the weeds and unwanted plants from going to seed. When you incorporate the cover crop, the shattered seed bank is dramatically

Table 10.15

	Block 1	Block 2	Block 3	Block 4
Year 1	**Hardy Crops** Brassicas, peas, some flowers, lettuce, year-2 biennials	**Cover Crop**	**Tender Crops** Tomatoes, peppers, cucurbits, beans, other flowers, corn	**Cover Crop**
Year 2	**Cover Crop**	**Hardy Crops** Brassicas, peas, some flowers, lettuce, year-2 biennials	**Cover Crop**	**Tender Crops** Tomatoes, peppers, cucurbits, beans, other flowers, corn
Year 3	**Tender Crops** Tomatoes, peppers, cucurbits, beans, other flowers, corn	**Cover Crop**	**Hardy Crops** Brassicas, peas, some flowers, lettuce, year-2 biennials	**Cover Crop**
Year 4	**Cover Crop**	**Tender Crops** Tomatoes, peppers, cucurbits, beans, other flowers, corn	**Cover Crop**	**Hardy Crops** Brassicas, peas, some flowers, lettuce, year-2 biennials

reduced. After a cover crop year, when you return to a seed crop, there will barely be any volunteers from that previous seed crop.

HOW TO EARN MORE REVENUE FROM YOUR SEED CROPS

At the beginning of this chapter, you figured out your growing space benchmark. As you planned out your seed crops, you were able to estimate how much revenue they could generate per area (In Col J of the "Crop Planner" tab.) Now, it's time to compare those numbers. If your estimate is lower than your benchmark, this crop is probably not going to be profitable for you to grow. If the estimate is higher, then the crop is probably going to be profitable.

I estimated that Tokyo Bekana generates $4.20/bed-foot. Let's compare that with the growing space benchmark from the two farms we calculated at the beginning of this chapter:

Farm A needed $5.74/bed-foot and Farm B needed $3.44/bed-foot.

Farmer Dan's $4.20/bed-foot looks good for Farm B but it is too low for Farm A.

How to Improve Your $/Bed-foot or $/Bed-meter

Here are some ways that you can increase the sales per growing space you can get from a seed crop.

You can negotiate a higher price with your seed contractor

Some seed companies are more responsive than others to haggling over contract prices. If you are a reliable grower and always produce top quality seed, you'll have more sway in these discussions. Growing less common varieties can also increase the price of your seed crop.

> *It would take $190/kg or $0.19/g ($86.36/lb or $5.4/ounce) for this Tokyo Bekana estimate to get to $5.70/bed-foot.*

You can improve your yields

For some crops (like brassica greens) this might mean harvesting a seed crop in a more timely manner so your seed doesn't shatter on the ground. For others it might mean improving soil fertility. However, when planning, it is always good to use conservative yields.

> *If I could increase my Tokyo Bekana yield to 41g/bed-foot, my sales estimate would be $5.74/bed-foot.*

You can harvest some of the fresh crop for market

Harvesting a couple of cuts from your leafy green seed crops, or some stems from your flower seed crops lets you generate more revenue from the same plants.

> *I get out my harvest knife and cut 0.4kg/bed-foot of small Tokyo Bekana leaves for my salad mix. At $20/kg ($9/lb) that comes to $8/bed-foot for the salad green alone. Adding the seed harvest, brings the sales estimate to $12.20/bed-foot.*

Chapter 11

Get Your Seeds Out Into the World

One thing you'll learn as you grow seed, is that plants are very generous. You will likely begin to accumulate more seed than you can ever use. I realized this pretty quickly when I harvested those first frozen tobacco berries way back in Chapter 1. From a handful of berries, I had tens of thousands of seeds. More than I would ever plant.

A couple of weeks after harvesting those berries, I went to the Montreal Seedy Saturday with a group of university friends. I packed up a dozen envelopes of tobacco seed for the seed swap table. In one of the urban farming workshops I started talking with another farmer named Emily who was at the event. She was interested in some of the tobacco seed but all the packs I'd brought were gone from the swap table. I told her I had some more tobacco seed that I hadn't brought that day and we exchanged phone numbers. Twenty years later, Emily and I are married and still going to Seedy Saturdays together. Of course these days we bring along Tourne-Sol seed racks and our kids are the ones scouring the seed swap tables.

As you accumulate a seed stash, it will be your turn to decide how to engage with your community through seed. Sharing with other farmers, gardeners, or local seed libraries is one option. You might run more organized seed exchanges with other farmers who also grow seed. And you might decide to sell seed.

This chapter focuses on selling seed as part of distributing seed into the world because that part is less intuitive and runs up against the real-world barriers of legality and money. But just because I don't put the same emphasis on sharing seed in this chapter, don't think it is any less important.

There are two main ways you might consider selling your seeds:

1. You can sell seed in bulk to a seed company that then sells it as small packs.
2. Or, you can sell seed in small packs yourself (essentially you become a seed company).

This chapter tackles the things you need to know for both ways to sell seed, but first ...

CONSIDERATIONS BEFORE SELLING SEED

Make Sure You are Selling Good Seed

Is this too obvious a point to lead with? It might be but I'm still going to say it. *You should only sell good seed.*

Good seed can mean a lot of things but in this case what I mean is:

- Good seed has good germination rates.
- Good seed is clean and has no debris (or very little debris).
- Good seed is true to type; that means the crops your seeds produce look the way the parents did.
- Good seed is disease free and pest free.

What kind of seed makes you happy when you shake it out of the pack? And keeps you happy when it grows into an adult plant? That's the kind of seed you want to share and sell.

Your seed company will be judged on the results your customers get from your seed.

Seeds and intellectual property

Plant Breeders Rights (PBR), Plant Variety Protection (PVP), and Utility patents are all different ways that seeds are treated as intellectual property. Depending where you are, certain seed varieties are considered intellectual property and you might not legally be allowed to sell them.

The Bauta Family Initiative on Canadian Seed Security has a great document that covers many of these topics called *Can I save it? A guide for seed savers on plant intellectual property rights in Canada.*

Consider where your seed comes from

Another thought before selling the seed you've grown is to consider the origins of your seed varieties. Are these varieties that have been stewarded or developed by farmers or plant breeders but have never been registered as intellectual property through legal means? Are these varieties that are distributed with Indigenous names or stories?

(In a larger context, only varieties with clear Indigenous origins are referred to as Indigenous varieties, but almost all crops ultimately have Indigenous origins. All or most of the major crops we grow wouldn't be here if it wasn't for the eons-long work of Indigenous seed-keepers.)

These varieties may be public domain seed so you can legally distribute them, but you might also want to recognize the seed's origins. At the least you can credit and mention where your seed varieties come from. You may also want to ask permission whether you can distribute this seed. You may also want to give back a portion of the sales to recognize the work that made this seed available in the first place. None of this may be required by law but it does make for good neighbors.

Reaching out to folks to talk about their seeds is also a great way to develop friendships in the seed world and make seed systems stronger and more resilient.

Is it Legal to Sell Seeds?

The answer depends on where your business operates. Here is some basic information on regulations in different countries.

> *But first, a disclaimer. I am not a lawyer or an expert on these topics. These are things I think you should be aware of but it is your responsibility to do your due diligence on how these apply to you and the varieties you work with. This is not legal advice!*

In Canada, seed sales are regulated at the federal level. For vegetable and flower seeds there are two main regulations:

- Your seeds need to exceed minimum germination rates.
- Your packs need to be labeled with either the germination rate and test date or with the intended year of sale.

For grain crops there are many more legal requirements.

The United States has similar federal laws but individual states also may have requirements as to whether you need a permit to sell seeds.

The European Union has a list of official varieties. Farmers can only grow varieties registered on that list and it is illegal to sell varieties that are not on that list. It is expensive and takes time to get a variety on this list, so only large seed companies do this. In recent years, there have been exceptions permitting sale of non-registered varieties to gardeners. Farms can still not legally purchase these seeds.

I am not familiar with seed regulations in the rest of the world, but I can only imagine they exist.

One note about seed regulations—producing, using, and distributing your own seeds without consideration for legal requirements is also a means of resistance and a political gesture. This can be a means for a community to assert their sovereignty. In these cases, not following the regulations is kind of the point.

SELLING BULK SEED TO SEED COMPANIES

This is probably the most accessible way to sell seed. You grow a bunch of seed and then you ship it to a seed company. They make sure it germinates well and then they send you a check.

Usually there is a step before this where you negotiate a seed contract but you can also just reach out to a seed company and offer seed that you have on hand. (They might say no.)

You might only grow one seed crop on your farm to sell in bulk or you might grow many seed crops to sell this way. This can easily piggyback into the existing spaces in a market garden and build to an interesting amount of seed and revenue.

How Tourne-Sol Started Selling Seeds on Contract

Back in January 2005 when we were getting ready for our first growing season at Tourne-Sol, I was once again at the Guelph Organic Conference. I went and talked to all the seed companies in the trade show. I asked them if they ever contracted seed from small farms.

Two small Ontario-based seed companies said they would have me grow seed for them. They each gave me seed packs for a few different tomatoes and told me they would buy a few ounces of seed if I could grow it out.

That fall I sent them the seed and made $600. That wasn't much compared to the 110 weekly vegetable baskets and the two farmers' markets we also did in year one. But it was still a decent return for the work that went into growing and cleaning the seed. And it was easy to double the next year as I grew more tomato seeds and added pepper seed on contract.

This amount got even bigger as I started to have surplus amounts from our own seed crops that those seed companies were interested in purchasing from us. Over the next seven years, we expanded our seed company clients until we were providing seed to a dozen seed companies.

What Companies can You Sell Seed to?

You can divide seed companies into two categories: small bioregional seed companies and larger seed companies.

Small bioregional seed companies are usually run by one or two people with a couple of employees. They focus on an offering of seeds adapted to their bioregion. Their sphere of sales tends to be in the states or provinces surrounding where they are based. They grow a large portion of the seeds they sell but very few small bioregional seed companies grow all the seeds they sell. The seeds they don't grow are either grown by contract growers in their bioregion or purchased from wholesale sources who source their seed all over the world.

Larger seed companies are bigger in every way than their small bioregional counterparts. They have many more employees. They sell to clients across the country. Many of these companies started out 20–30 years ago as small bioregional seed companies and have grown to meet a large seed demand. Though a few larger seed companies grow a portion of their seed offering, the bulk of their seed comes from wholesale sources and to some extent contract growers.

Both small bioregional seed companies and large national seed companies are looking for seed growers. And to some extent their seed grower needs are similar. Both types of seed companies want quality seed—seed with good germination rates that is disease free and grows true to type. But the big difference is the volume of seed that these companies need. Large national seed companies distribute hundreds (if not thousands) of pounds of seeds of some varieties whereas small bioregional seed companies distribute a fraction of that. This means that it can be difficult to meet the needs of a larger seed company when you begin to produce seed. It also means that the price per pound of seed that a large seed company will pay is probably smaller than what a smaller company will pay.

If you want to go even deeper into seed contracts, you should go look at the Organic Seed Alliance's Seed Economics Toolkit. (https://seedalliance.org/publications/seed-economics-toolkit/).

The Seed Worker Organizing group (https://seedworkers.org/) has been working on seed production contract guidelines to foster mutually equitable contracts and working relationships between seed producers and seed sellers. This is done by a volunteer working group who work in an iterative manner based on feedback from the seed community.

With this in mind, market growers who want to break into selling seed to seed companies should focus their attention on small bioregional seed companies. Once they have increased their production capacity and developed their expertise, they might then wish to approach larger seed companies to be contract growers.

Finding Seed Contracts

A seed contract is an agreement between a grower and a seed company for how much seed of a specific variety the company will purchase. It also specifies the price that will be paid, and considerations like germination rates and isolation distance. The contract might also include stipulations related to what happens to any overproduced seed.

The first step to get a seed contract is to get in touch with seed companies. This can be as simple as sending an email, or you can go to a farm conference or a Seedy Saturday where seed companies have booths. Wait till there is a lull in the client flow and go introduce yourself to the vendors and tell them you're looking to grow seed on contract.

You should come prepared with a few answers.

1. Are you certified organic? If the company only sells certified seed, this is a deal breaker. For companies that sell organic and conventional seed, they might pay a premium for organic seed.
2. Have you ever grown seed before? And if so, what crop species and in what amounts? This immediately gives the seed company an idea of what you might be capable of.
3. What kind of isolation distances can you provide? Keeping crops from cross-pollinating is vital.

What Seed Crops Will You Grow?

When you first approach a seed company, you should be clear what seed crops you have experience with. These are likely the crops that the seed company will be most comfortable contracting you for, though they might ask you to grow varieties you haven't grown before.

The seed company will provide you the seed and you'll grow it out. Be cautious when committing to varieties you haven't grown before. They might not yield as much as other varieties of the same crop that you have experience with.

In most initial seed contracts you'll be growing out crops that the company needs. However, if you have good varieties that you've been maintaining or have developed, seed companies might be interested in these too. If they are interested, they'll ask you to give them a seed sample of your variety so they can trial to see if it is a good crop for their market. Then they might buy seed you already have on hand, or they might contract you to grow a lot for them.

Types of Seed Contracts

There are two basic frameworks I've seen around seed contracts.

In most cases, a contract is for a specific weight of seed at a specific price.

There are also some seed companies that pay their growers a portion of the seed pack price. Often 30–50 percent of the price. So If a seed pack is $5, then the seed company would send $2.50 to the grower. In some cases the company sends the grower the seed envelopes, and the grower then packs them with the appropriate seed per pack. They then send the packed seed to the company.

From a financial perspective, getting a fixed percentage of each pack can be very appealing, since you make more money per pound of seed you grow. However, you also only get paid as a function of the seed envelopes the company sells. It may take a few years for the seed to sell.

A seed company that has contracted a set amount will pay for the seed as soon as it passes germination testing, and then they assume the risk of selling the seed.

In some cases, seed companies will also pay their growers in installments during the duration of the contract. At Tourne-Sol, we pay our regular growers in three installments: we pay one-third when the crop is planted, one-third when the crop is harvested based on the forecasted amount of what the grower expects to deliver, and we pay the balance once the seed has passed the final germ test.

Evaluating Seed Contract Prices

Seed companies usually have their own standard rates for specific crop types or specific varieties. You can use the Seed Farmer Crop Plan spreadsheets from Chapter 10 to crunch your seed contract prices and evaluate how profitable they are for your farm. You'll see that price per kilogram or pound makes a big difference in determining which crops can be profitable for you.

You'll also see that seed quantity also has an impact: the greater the amount of seed sold of one variety, the lower the price paid per kilogram or pound. Smaller seed contracts will usually have a higher dollar per kilogram or pound amount.

Generally biennial seeds and uncommon varieties get higher prices per kilogram or pound than annuals and common varieties.

If you're set up to produce a lot of seed, you might find that you can still be profitable with a lower dollar per kilogram or pound. But if you have limited space, you will need a higher ratio to warrant taking a seed contract.

Make sure to be clear with your seed companies whether you can afford to accept their prices. You might find there is some room for negotiation. Most seed companies want their growers to be able to run profitable businesses and be able to keep supplying them with good seed.

Responsibilities of Being a Contract Seed Grower

When you take on a seed contract, you want to do a great job to impress that seed company. This is how you show them that you are reliable and can take on more contracts.

The first thing of course is to send them great seed but there are some other things you can do to make them love working with you.

Keep companies informed during the growing season

You should keep your seed company up to date as to how your seed crops are growing.

There may be specific checkpoints in the contract, but even if there are not, it is good practice to send out the occasional email to your seed company to tell them that the crops are in the ground and how they are growing. This helps create a good relationship with your client and will be one reason that they wish to keep working with you (assuming you also deliver good seed!).

It is especially important to tell a seed company when you have unexpected problems and don't think you'll be able to deliver the right quantity of seed, or that there may be delays in seed cleaning. You should also tell them if there are potential disease problems or unexpected cross-pollination.

The sooner a seed company gets this information, the more they are able to adjust their own sales plans to compensate for a seed shortfall, and the sooner they can work with you to mitigate any problems.

Deliver seed on time

This is the next most important thing. Once a seed company has the seed, it still might need to do some final seed cleaning and get some germination tests done. Only then will they be able to sell it.

It is much easier for the company to fit that into their normal schedule then to have to deal with late seed from you.

Pack your seed so it doesn't spill

When you deliver your seed at the end of the season, put it into thick bags. These could be feed sacks or zippered plastic bags. For zippered bags, you should double bag the seed as an extra precaution to avoid spilling. Make sure to label your seed clearly. You should also provide a pack list of what varieties and quantities are in the shipment.

Pack boxes of seed so that seed bags don't move around. You can pad with newspaper or magazine pages, or recycled packing materials from your own online shopping. You should also avoid overpacking boxes too. This can result in too much force on the seed when mail couriers are handling your package.

If you ship the seed in the mail or by courier, send any tracking numbers to your seed clients.

Build Relationships

Initially a seed company is taking a risk by working with you. They don't know if you'll deliver on time or in the right quantity. And they won't even know if your seed is disease-free or true to type until the following season. This can make working with new growers really scary for a seed company.

But as you deliver good seeds on time, you build a reputation that you are a reliable business associate, so they are increasingly excited to work with you, especially if you also communicate in a timely manner.

As your relationship builds, they will be willing to trust you with increasing amounts of seeds and numbers of varieties. And you might find you can grow a lot of seed on contract!

Now let's talk about what it's like to be on the other side of the equation and sell seed as a seed company ...

SELLING SEED PACKS TO GARDENERS AND FARMERS

Selling seeds in smaller packs directly to gardeners and farmers is another way to sell your seeds. This essentially means running your own seed company.

Should You Start a Seed Company?

I want to start this section with a bit of tough love.

Selling seed in small packs directly to consumers can seem like a no-brainer for those of you who like to crunch numbers quickly. You can sell a pound of tomato seeds for $400 or you can break that same pound into 2,000 packs and sell each pack for $4 to make $8,000!

Isn't $8,000 of sales better than $400 for the same pound of tomatoes? If only it were that simple.

To start, this math is only profitable if you can actually sell 2,000 packs of the same variety of tomato. This math also disregards all the work it takes to get the seeds into those little envelopes and then into your customers' hands.

So I'm very hesitant to recommend that you start your own seed company. But I know that some of you reading this book still want to start a seed company and I want to make sure you think through all the parts of what you're getting into so that you can be successful.

The first thing you should do is ask ...

Why do you want to sell seeds?

Is it as simple as wanting to generate more revenue from your farm? If that is the case, you should definitely run the numbers on how a seed company enterprise on your farm compares with the way you currently grow and sell crops from your farm. Maybe you should just expand your veggie basket program, or go to another farmers' market instead of diversifying your business. Maybe you should simply eliminate your current less-profitable crops and optimize what you're already doing.

Do you want something that is easier than growing vegetables and lugging them to market? Seeds are definitely lighter to lug around than bins of cabbage.

Has your seed-saving gotten out of hand and you have more seed in your cupboards than you know what to do with? These can all seem like logical reasons to want to diversify and change your farm offerings to include seed, but these reasons on their own might not get you through the tough times that come with running any business.

You need a deeper motivation pushing you to keep going at this.

Are you fascinated with the stories behind the seeds? Are they your ancestral crops that you want to keep in circulation? Are there varieties and whole crop families that you can't find in seed catalogs and you know you're not the only one, so you want to make these available to a larger public to make sure farmers can grow crops that folks want to eat? Do you simply want to make sure the range of diversity doesn't get lost? Do you have new varieties that you've been selecting on your farm that you want to get into the hands of other farmers?

These are motivations that are more likely to let you tough it out when the realities of running a business set in. Because running a seed company is a business, and if you're going to be successful, you need to treat it as a business and understand your numbers.

That being said, you can dabble your toes into the seed company waters without much risk. You can start small and see how you like it and slowly expand as the market develops. This is especially true if you already have a farm business that is covering your expenses and providing you with an income.

The Seasonal Workload of Growing Vegetables/Flowers and Selling Seeds

There is one other thing to consider if you're a farmer thinking of adding a seed company to your farm. And that is when the work happens.

Have you thought about your seasonal workload?

Growing market crops and growing seeds are very similar activities: You plant and weed and harvest and pack and get them to your clients. But these activities don't happen the same way for market crops as they do for seeds. That has a huge impact on seasonal workloads.

You grow, harvest, and sell vegetables at the same time of year. That makes for intense vegetable summers.

But with seeds, the growing and harvesting happens in one season and the selling happens in another season.

Yes, summer is definitely busy for the seed crew. But winter is when things get serious. The orders just keep coming in and you've got to make sure they ship in a timely fashion.

In the table on the next page I've sketched out what that looks like for a farmer in a northern climate.

Yes, this is a dramatic oversimplification.

The key thing to notice is that you work continuously through the year and don't have tons of time to recuperate and get ready for the next season.

I've repeated my warnings a number of times that running a seed company is not easy. And ultimately what I don't want to see is potential seed growers burn out and quit running their business and quit growing seed. I'd rather see folks dabble with seed and slowly get into distribution in a way that makes sense.

All right, so let's put my seed company warnings aside and jump into what it means to get a seed company started!

Table 11.1

Month	Market Vegetable Grower	Seed Grower with own Retail Seed Company
January	Time to rest	Things are getting busy (orders start coming in)
February	Time to rest	Peak Activity (orders, orders, and orders)
March	Excitement! (starting greenhouse)	Peak Activity (orders, orders, and orders)
April	Excitement! (we get to go outside)	Not quite as busy! (orders are slowing down)
May	The race begins (get those plants in the ground)	Wrapping down (get to play outside and plant)
June	The race is on (get those plants weeded)	Time to rest
July	Things are getting busy (main crops coming in + sales)	Time to rest
August	Peak Activity (big harvests and sales)	Excitement! (seed harvest is on)
September	Peak Activity (big harvests and sales)	Excitement! (starting to clean seed)
October	Not quite as busy! (harvest is slowing down)	The race begins (race against weather to harvest)
November	The race is almost over (clean up that field)	The race is on (cleaning and germ tests and packing seeds!)
December	Wrapping Down (get to plan next year)	The race is still on (pack those seeds)

Ways to Sell Seed Packs

How Tourne-Sol started selling seeds in seed packs

January 2007 is when we started selling seeds directly to gardeners. I wrote up a seed catalog with 20 varieties listed in it. We went to the Montreal Seedy Saturday and had a booth. We got a handful of mail orders. Over the next couple of years we added a few more Seedy Saturdays and our mail orders increased. In 2011, we set up an online store to more easily take orders. During this time the portion of our revenue from seed increased every year, but it was only 10-20 percent of our gross sales. Our veggie baskets and farmers' markets accounted for the bulk of the farm sales.

In 2016 we started to put more resources into growing our seed company. By 2022, about 50 percent of our revenue comes from seed. During the winter, we have ten people working part- and full-time to sell, pack, and ship seeds from our seed warehouse. During the summer this team is down to about four people. The rest of the crew moves into growing and harvesting vegetables.

It took us over ten years to get to this point. And I'm glad we were able to take it slowly because we could rely on vegetable sales. If you are thinking of moving into seed sales, it's worth having other farm enterprises to support you. The longer you can hold off needing most of your income coming from seed, the more time you have to hone your seed skills and marketing skills and to build relationships.

Farmers' markets

If you are already at a farmers' market, then adding seed packs as one more market product is very easy. You can pack up some seed and put them in a small box with dividers or in a small seed rack and have it by the checkout area of your market stall.

This is especially good if your markets start early enough and overlap with when folks are putting in their gardens. If you're already selling seedlings in three-inch pots, seed packs are a natural addition to tap into a market you already have.

You can keep selling seed through the season, even after folks traditionally plant their gardens by telling folks about succession planting and any seasonal plantings they might not know about, such as fall gardening.

You don't need a big inventory of packed seeds, you just need to keep them dry during your markets. They are non-perishable so you can bring back unsold packs week after week.

This is the least intensive and least risky way to start selling seed packs.

It can give you a feeling for what is involved.

Farming and gardener events

These are events where you physically go and set up a booth with your seed packs. Customers come and peruse your stuff. The strength of these events is that you get to meet face-to-face with customers who are specifically looking for seeds. You can talk to them directly to tell them about the stories behind your farm and your seed.

If you go to the same events year after year, you'll start to recognize faces and build up a group of regulars.

There is a rush to have your seeds packed for the event but otherwise the main work is attending it. You might get stuck with unsold seed packs but initially it will not be a ridiculous amount.

You can simply go to a couple of these and sell a bunch of seeds. Or you might find a number of events like this through the winter and spring where you can move an interesting number of packs.

There is a range of different types of events where you can sell seeds.

Community seed swaps and seed sales

Some communities have annual seed swaps and seed sales where local seed folks can go and sell their seed. Most of these events also feature workshops and talks. They are go-to events for all of the gardeners in the area. Some of these events only have seed swaps, but others invite small vendors to set up booths.

These events are great because folks are there specifically looking for seed.

These events are especially well developed in Canada and called Seedy Saturdays and Sundays—once a year seed markets and seed swaps that pop up in communities across the country. There are over a hundred Seedy Days across Canada—Seeds of Diversity Canada has a listing of all the Seedy Days across Canada on their website. The smallest Seedy Days might feature one or two seed companies and a lot of swapping between local gardeners in addition to many non-seed businesses selling stuff like gardening supplies or artisanal products or winter vegetables or baked stuff. The largest Seedy Days can feature an overwhelming 20 to 30 seed companies and thousands of folks coming to do their annual seed shopping

Flower shows, farm conferences, garden talks

These are all events where you can also set up a booth and sell seeds. Not all customers are there looking for seeds, so you may have to work harder than if you were at a seed sale. But these are definitely folks who love seeds, so you might not have to work too hard!

The tricky thing with some of these is the registration fees. If the cost is too high it might quickly eat up any profit you can make. Expensive events can be good events once you do have a product you know will sell, and folks who are excited to find you.

Online store

This is where I feel I should warn you to be careful about your seed company dreams.

It has become very easy to set up an online store. Shopify, Woocommerce, or any other number of online stores offer you easy functionality to set up products and take payments. Some of these systems also integrate well with mail carriers. Technically, you can have an online store up in an afternoon. Practically, it takes longer than that—you'll also need some time to write your product descriptions and get pictures formatted to be online.

Getting products online is one thing, getting people to your store is another. Running a successful online store requires building a brand, social media presence, and newsletter. These things all take time. More time than setting up that store.

If your sales are low, you can pack seeds as the orders come in. But it doesn't take much volume in orders to make custom packing overlap with shipping orders later. In that case, you'll probably want to pack what you think will sell over the coming months in advance of shipping dates, so it is easy to fill orders.

Make sure to prioritize customer services. Reply to emails promptly and ship seeds even more promptly. If you can get orders out within a business day of receiving them, folks will notice.

It is very easy to set up an online seed store—maybe a little too easy. I have heard from a number of folks who threw themselves into an online store that they found it was harder to sell enough seed to generate the revenue they wanted and a lot less profitable than they guessed it would be.

What about print seed catalogs?

You probably love receiving your favorite seed catalogs and perusing them on the kitchen table, dreaming about all the choices.

Still, you probably shouldn't start with a print seed catalog these days. They take time to design, cost money to print, and cost money to ship. For a new seed company it is difficult to recoup the cost.

Online stores offer most of the same benefits for probably less investment.

You can always go a simple route in black and white and a few pages. Or have something folks can download as a PDF. Or that you can print and send to those folks who request a print catalog.

If you hit a certain sales volume and you still dream about catalogs, then maybe go for it. But I would be surprised if in the current age of online marketing, you can't just go further with newsletters and social media.

Seed racks

Seed racks are a way to sell seed packs wholesale. A garden center or food co-op or some other store buys seeds from you. They display them on a seed rack in their business for consumers to purchase.

The strongest part of this relationship is the volume of packs they will initially take. If your seed rack has 48 pockets and you put ten packs per pocket that is 480 packs sold per rack order. That is the equivalent of many online store orders.

However, seed rack programs usually buy seeds at a discount. They will usually pay you 30 to 50 percent less than the price they will sell the seed packs at. This can still be valuable because you don't have to work as much for those orders.

You also need to decide whether to offer buy-back or consignment for your seed rack program. Can stores return a certain number of unsold packs for a refund or credit on next year's orders? If they can, this adds an element of risk since you don't know what your actual sales are until the end of the season.

There is some vulnerability to relying too much on seed racks. Your seed vendors might decide to work with someone else the following year, or if their sales are down they might decide to reduce their order. This can result in dramatic swings in sales from one year to the next.

Some other challenges that come along with seed racks include needing good invoice systems that can handle a large number of SKUs and sync with your online store. Tracking down late payments takes time and can be stressful on cash flow. Customer service is really important—rack customers often order pack refills exactly when they need them so you need to get those seeds out the door ASAP.

Another challenge is that once your seed packs are out of your hands, it is hard to control quality. Rack customers might be displaying seeds in an overheated hot greenhouse, or might be keeping leftover seed from year to year, which might mean they are selling low-germ seed. Your reputation relies on this seed!

You probably shouldn't start off by running a full wholesale seed rack program right out of the gate. A few seed racks in local businesses that you can visit can be a good way to get into seed racks while keeping the risks at bay.

Retail Seed Packs Considerations

The other part of selling seed retails is your product! Your seed packs.

Here are some of the main decisions you'll need to figure out to get a marketable product.

What Will Your Packet Look Like?

Print your own labels. You can start pretty DIY. You can get some kraft coin envelopes and then print some labels that you stick to each pack. This might seem a little basic but it can go a long way to get started.

If you are selling face-to-face with clients, your sales pitch and verbal banter is probably the biggest thing that will influence someone to buy your seeds. If you're selling online, folks don't see what your packs look like until they get the order. If every other part of their seed experience with you is awesome, they'll come back next year for more, no matter what your seed packs look like.

Print your own packs. If you want to get a little more sophisticated in your DIY, you can print information directly on your packs instead of using labels. Many home printers will work for small print runs. But as your print runs get larger, home printers might not be fast enough and might degrade in quality as the printer gets older. You can consider commercial printers but they will cost anywhere from a few thousand dollars to over ten thousand.

You can also get an envelope supplier to print some of your basic information on envelopes—your logo and farm name, address, and maybe a little blurb. All the information that is the same across all your packs. Then you just print the information that is unique to each variety on these packs.

Buy fully printed packs. The other option is to design seed packs for each variety with color photos or drawings and all the associated information.

These can look really slick and make it much easier to sell your seeds when you're not there, which is an important consideration for seed racks. But you often need to buy seed packs in lots of a thousand envelopes of the same variety. You might not yet be at a scale where you can sell that many seed packs of most of your varieties, even over a few years.

One challenge with fully printed packs is how to add information for specific lots on the seed pack. You can use a price gun or a label to do this. You can also leave a blank part on the envelope where this information is printed.

We started off with printed labels and made our way up to ordering fully printed packs. When we received them they looked great and they made our seed rack look fantastic.

But one thing surprised me—they didn't necessarily increase sales at our in-person events such as Seedy Saturdays and farmers' markets. At those events, folks were coming to get seeds and were going to buy them one way or the other. If anything, fancy seed envelopes left a negative impression on some clients who felt that we had gone too commercial.

You don't need the fanciest packs to develop a dedicated following.

What information do you need on your seed packs?

You should check your local regulations for what information is required on your seed packs. This is what you probably want to include:

- Crop type
- Crop variety
- Seeds or weight per pack
- Company name
- Company contact info
- Seed lot number
- Germination results and germination date, or the intended year of sale
- Variety description
- Growing information

Filling your seed packs

You're going to need to figure out a way to make sure you have roughly the same amount of seeds in each of your seed envelopes. Most small companies sell seed by weight in grams or in fractions of ounces. But they don't actually weigh each pack. Instead they use calibrated scoops to fill the packs.

You'll need to have a collection of scoops to be able to meet a range of weights.

A collection of kitchen measuring spoons is a good starting place— 1 tbs, ½ tbs, 1 tsp, ½ tsp, ¼ tsp are all valuable. You also want to have pinch, dash, and smidgen sizes. This will give you a broad range of seed-packing options.

The next step to expand your scoop range is to get some gunpowder scoops. These have many sizes under a teaspoon.

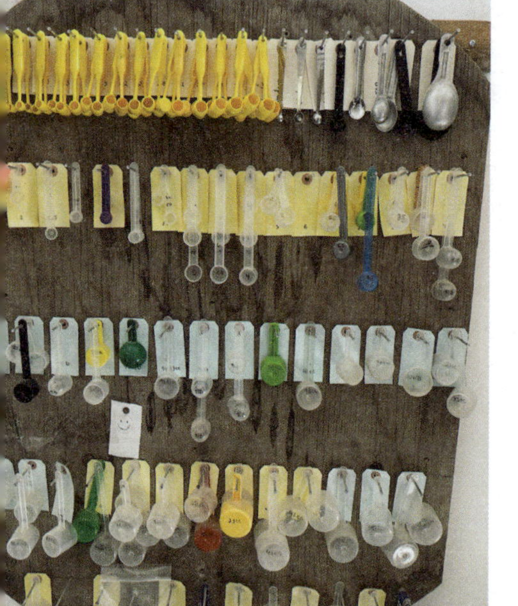

Fig. 11.1: *A wall of seed scoops for all your packing needs.*

To calibrate your seed scoops:

1. Check how much seed you want in a pack.
2. Choose a scoop that you think will be the right size.
3. Measure ten scoops of seed into a container.
4. Weigh the container contents.
5. Divide the weight by ten to get the weight per scoop.
6. If this weight per target scoop is your target weight then you have found your scoop. Yay.
7. Otherwise, repeat these steps with another scoop size.

Note down what scoop you used to pack a seed lot so you have a good starting point for your next pack run. It is a good idea to calibrate your scoop each time you pack a lot to make sure you have the right scoop.

How much seed should be in each envelope?

Each seed company makes their own decisions on this. But here are some guidelines:

- Put more small seeds in a pack, for example, lettuce.
- Pack fewer large seeds such as beans.
- Put more seeds that are sown in the garden, such as carrots.
- Fewer seeds that are started indoors, such as tomatoes.

Look at other seed catalogs to see how many seeds they include per pack.

You'll also want to consider the dollar value of the seed in your pack based on bulk prices.

Fig. 11.2: *Packing seed packs one scoop at a time.*

Does the world need another seed company?

There are some really great seed companies out there. They might be what is inspiring you to want to start a seed company. But what I'm trying to highlight over and over is that starting a seed company is not easy work and isn't a get-rich-quick scheme for a small farm. If you're simply going to start another company that offers similar varieties to what is out there, then you might find that the world doesn't really need another seed company.

But if you have seeds and stories that are different from what is currently on offer, if you have seeds that are better adapted to your bioregion and our increasingly extreme weather of all kinds, if you are supporting your local communities and building their seed resilience and food sovereignty, and if you can run this business in a way that is sustainable to you and that keeps you doing this for years—then yes; I think the world does need another seed company and that you might have a real gift to offer the world.

Afterword

ALL RIGHT, so what is left to say as we get closer to farewell?

You've already chosen a couple of crops that you will let go to seed. You know not to worry about cross-pollination or perfectly clean seed or getting 100 percent germination rates or any of those spreadsheets. Your only goal is to harvest that seed and then use it next year. That's all you need to do to start your seed farmer journey.

Now it's time to head out to the field. Put down this book (or maybe shove it in your backpack for further consultation) and grab some flagging tape and a couple of empty envelopes and go see what you can find.

If it's peak summer there will be ripe tomatoes or peppers or some forgotten calendula blooms that have turned to dried seed. If it's a different time of year, there are still things to find. It could be some solitary nigella seed pods sticking out of a snow drift. Harvest a pod or two and bring it home. Or it could be some chard plants that survived until spring, or even sprouting turnips and onions in your cold room. You could replant these to see what happens.

I started this book by telling you that **you can grow seeds on your farm. It's easier than you think.**

This is only true because there is already seed growing on your farm.

All you have to do is notice this potential and let your plants go to seed.

Dan
September 2024

Appendix 1 Seed Farmer Spreadsheets

You can head to sheets.seedfarmerbook.com to get the Seed Farmer spreadsheets in the book along with a series of videos to guide you through how to use each sheet.

Seed Order Spreadsheets

These are the sheets used in Chapter 1 to compile and analyse your seed order to help you decide what crop varieties you might grow for seed.

Germination Test Spreadsheets

These are the sheets used in Chapter 8 to record your germination tests. There is also a summary sheet to give you an overview of your seed lots and their germination status.

Crop Plan Spreadsheet

These are the sheets used in Chapter 10 to plan out how much you should grow of each crop to get the seed you need. These sheets also generate field and nursery schedules to guide you through your season.

Appendix 2 Glossary

HERE IS A PARTIAL LIST of terms that come up throughout this book.

Crosser: These are plants that predominantly cross-pollinate.

Decant: A seed-cleaning technique where you put seed in a container and fill the container with water. In most cases the viable seed will sink and hollow seed and other plant material will float. You can remove the unwanted floating bits by pouring off the water and leave the heavier seed at the bottom of the container.

Hardy crops: Crops that will survive light frosts. Some hardy crops can survive some pretty heavy frosts.

Hybrid seeds (also abbreviated to F1 seeds): These are created by crossing two plant lines and are marked on seed packs and in seed catalogs with an F1 in their name. The seed you sow will be very uniform but if you keep seeds from a hybrid and grow them out you will get a large diversity of different characteristics.

Isolation distance: The distance between two crops that might cross-pollinate together.

Off-type: A plant that is different from the others in a plant population. If you want to maintain uniformity in your varieties, you should remove off-types. However off-types are also exciting opportunities to create new varieties in your fields.

Open-pollinated seeds (also abbreviated to OP seeds): Stable varieties that will give you the same variety the next year as long as you control any unwanted cross-pollination.

Pappus: The feathery parachutes attached to the seeds of some crops in the Asteraceae family such as dandelions.

Selfer: Plants that are predominantly pollinated by themselves. These crops barely cross-pollinate at all. Selfers have hermaphroditic flowers that are wrapped in tight petals that make it hard for insects to get to the pollen, and even harder for them to spread that pollen to other plants.

Siliques: The seed pods on brassica plants.

Tender crops: Crops that won't survive a light frost.

True to type: An expression describing when the plants of a seed variety look the way the plants in the previous generation looked.

Appendix 3 Resources for Further Seedy Study

Books about Growing Seed

How to Save Your Own Seeds: A Handbook for Home Seed Production, 6th Edition (Seeds of Diversity Canada, 2013).

The Organic Seed Grower: A Farmer's Guide to Vegetable Seed Production by John Navazio (Chelsea Green, 2012).

The Wisdom of Plant Heritage: Organic Seed Production and Saving by Bryan Connolly (NOFA-NY, 2004).

The Seed Garden: The Art and Practice of Seed Saving Contributions by Jared Zystro, Micaela Colley; edited by Lee Buttala, Shanyn Siegel (Seed Savers Exchange, 2015).

Seed to Seed: Seed Saving and Growing Techniques for Vegetable Gardeners, 2nd Edition by Suzanne Ashworth (Seed Savers Exchange, 2002).

Small-Scale Organic Seed Production by Patrick Steiner (Farm Folk City Folk). (You can download a free PDF from https://farmfolkcityfolk.ca).

Books about On-farm Plant Breeding

Breeding Organic Vegetables: A Step-by-Step Guide For Growers by Rowen White and Bryan Connolly (NOFA-NY, 2011).

Breed Your Own Vegetable Varieties: The Gardener's and Farmer's Guide to Plant Breeding and Seed Saving, 2nd Edition by Carol Deppe (Chelsea Green, 2000).

Landrace Gardening by Joseph Lofthouse (Father of Peace Ministry, 2021).

Can I save it? A guide for seed savers on plant intellectual property rights in Canada by Christy Ó Ceallaigh-Bisson, Kaitlyn Duthie-Kannikkatt, and Aabir Dey, (The Bauta Family Initiative on Canadian Seed Security, 2022).

Seedy Podcasts

SeadHeads, hosted by Steph Benoit and Hugo Martorell.

Seeds & Their People, hosted by Chris Bolden-Newsome and Owen Taylor.

Free The Seed!, hosted by Rachel Hultengren.

Seed Farmer, hosted by Dan Brisebois (that's me!).

Seedy Organizations

Seeds of Diversity Canada https://seeds.ca/

Bauta Family Initiative on Canadian Seed Security https://www.seedsecurity.ca/en/

Organic Seed Alliance https://seedalliance.org/

The Open Source Seed Initiative https://osseeds.org/

Appendix 4 Seed Reference Chart

Source document:
The Seed Farmer Appendix 3—Shared

This is also available as a spreadsheet at sheets.seedfarmerbook.com

Seed Reference Chart

Species	Crop	Selfer/ Crosser	Life Cycle	Type of seed	Germ tempera-ture	Yield g/bed-foot	Yield oz/ bed-foot	Yield g/bed-meter
Coriandrum sativum	Cilantro	Crosser	Hardy Annual	Naked Dry Seeds	60°F (15°C)	45 to 91	1.6 to 3.3	148 to 298
Spinacia oleracea	Spinach	Crosser	Hardy Annual or Plant Biennial	Naked Dry Seeds	60°F (15°C)	27 to 64	1 to 2.3	89 to 210
Allium cepa	Onions	Crosser	Root Biennial	Naked Dry Seeds	68°F (20°C)	18 to 32	0.6 to 1.1	59 to 105
Calendula officinalis	Calendula	Crosser	Hardy Annual	Naked Dry Seeds	68°F (20°C)	18 to 32	0.6 to 1.1	59 to 105
Lactuca sativa	Lettuce	Selfer	Hardy Annual	Naked Dry Seeds	68°F (20°C)	9 to 27	0.3 to 1	30 to 89
Xerochrysum bracteatum	Strawflower	Crosser	Tender Annual	Naked Dry Seeds	68°F (20°C)	9 to 18	0.3 to 0.6	30 to 59
Amaranthus sp.	Amaranth	Crosser	Tender Annual	Naked Dry Seeds	86°F/68°F (30°C/ 20°C)	23 to 91	0.8 to 3.3	75 to 298

Seed Reference Chart

Species	Crop	Selfer/ Crosser	Life Cycle	Type of seed	Germ tempera-ture	Yield g/bed-foot	Yield oz/ bed-foot	Yield g/bed-meter
Anethum graveolens	Dill	Crosser	Hardy Annual	Naked Dry Seeds	86°F/68°F (30°C/ 20°C)	45 to 91	1.6 to 3.3	148 to 298
Beta vulgaris	Chard	Crosser	Plant Biennial	Naked Dry Seeds	86°F/68°F (30°C/ 20°C)	68 to 114	2.4 to 4.1	223 to 374
Beta vulgaris	Beets	Crosser	Root Biennial	Naked Dry Seeds	86°F/68°F (30°C/ 20°C)	68 to 114	2.4 to 4.1	223 to 374
Celosia argentea	Celosia	Intermedi-ate Selfer	Tender Annual	Naked Dry Seeds	86°F/68°F (30°C/ 20°C)	14 to 45	0.5 to 1.6	46 to 148
Cichorium endivia	Escarole and endives	Selfer	Hardy Annual	Naked Dry Seeds	86°F/68°F (30°C/ 20°C)	9 to 27	0.3 to 1	30 to 89
Cichorium intybus	Radicchio and chiories	Crosser	Plant Biennial	Naked Dry Seeds	86°F/68°F (30°C/ 20°C)	9 to 27	0.3 to 1	30 to 89
Daucus carota	Carrots	Crosser	Root Biennial	Naked Dry Seeds	86°F/68°F (30°C/ 20°C)	9 to 18	0.3 to 0.6	30 to 59
Helianthus annuus	Sunflower	Crosser	Tender Annual	Naked Dry Seeds	86°F/68°F (30°C/ 20°C)	23 to 91	0.8 to 3.3	75 to 298
Zinnia elegans	Zinnia	Crosser	Tender Annual	Naked Dry Seeds	86°F/68°F (30°C/ 20°C)	14 to 27	0.5 to 1	46 to 89
Nigella damascena	NIgella	Crosser	Hardy Annual	Protected Dry Seeds	60°F (15°C)	23 to 68	0.8 to 2.4	75 to 223
Papaver somniferum	Poppy	Selfer	Hardy Annual	Protected Dry Seeds	60°F (15°C)	14 to 27	0.5 to 1	46 to 89
Pisum sativum	Peas	Selfer	Hardy Annual	Protected Dry Seeds	68°F (20°C)	68 to 91	2.4 to 3.3	223 to 298

Seed Reference Chart

Species	Crop	Selfer/ Crosser	Life Cycle	Type of seed	Germ tempera-ture	Yield g/bed-foot	Yield oz/ bed-foot	Yield g/bed-meter
Abelmoschus esculentus	Okra	Intermedi-ate Selfer	Tender Annual	Protected Dry Seeds	86°F/68°F (30°C/ 20°C)	45 to 136	1.6 to 4.9	148 to 446
Brassica juncea	Mustard	Crosser	Hardy Annual or Plant Biennial	Protected Dry Seeds	86°F/68°F (30°C/ 20°C)	23 to 45	0.8 to 1.6	75 to 148
Brassica napus	Rutabaga	Crosser	Root Biennial	Protected Dry Seeds	86°F/68°F (30°C/ 20°C)	23 to 68	0.8 to 2.4	75 to 223
Brassica oleracea	Collards	Crosser	Plant Biennial	Protected Dry Seeds	86°F/68°F (30°C/ 20°C)	23 to 68	0.8 to 2.4	75 to 223
Brassica oleracea	Kale	Crosser	Plant Biennial	Protected Dry Seeds	86°F/68°F (30°C/ 20°C)	23 to 68	0.8 to 2.4	75 to 223
Brassica rapa	Brassica greens	Crosser	Hardy Annual or Plant Biennial	Protected Dry Seeds	86°F/68°F (30°C/ 20°C)	23 to 45	0.8 to 1.6	75 to 148
Brassica rapa	Turnips	Crosser	Root Biennial	Protected Dry Seeds	86°F/68°F (30°C/ 20°C)	23 to 45	0.8 to 1.6	75 to 148
Eruca sativa	Arugula	Crosser	Hardy Annual	Protected Dry Seeds	86°F/68°F (30°C/ 20°C)	23 to 45	0.8 to 1.6	75 to 148
Phaseolus vulgaris	Beans Bush	Selfer	Tender Annual	Protected Dry Seeds	86°F/68°F (30°C/ 20°C)	68 to 91	2.4 to 3.3	223 to 298
Phaseolus vulgaris	Beans Pole	Selfer	Tender Annual	Protected Dry Seeds	86°F/68°F (30°C/ 20°C)	68 to 91	2.4 to 3.3	223 to 298
Raphanus sativus	Radish, Spring	Crosser	Hardy Annual	Protected Dry Seeds	86°F/68°F (30°C/ 20°C)	23 to 45	0.8 to 1.6	75 to 148
Raphanus sativus	Radish, Winter	Crosser	Root Biennial	Protected Dry Seeds	86°F/68°F (30°C/ 20°C)	23 to 45	0.8 to 1.6	75 to 148

Seed Reference Chart

Species	Crop	Selfer/ Crosser	Life Cycle	Type of seed	Germ tempera-ture	Yield g/bed-foot	Yield oz/ bed-foot	Yield g/bed-meter
Zea mays	Corn, sweet	Crosser	Tender Annual	Protected Dry Seeds	86°F/68°F (30°C/ 20°C)	45 to 68	1.6 to 2.4	148 to 223
Capsicum spp.	Peppers	Intermedi-ate Selfer	Tender Annual	Protected Wet Seeds	86°F/68°F (30°C/ 20°C)	5 to 18	0.2 to 0.6	16 to 59
Citrullus lanatus	Watermelon	Crosser	Tender Annual	Protected Wet Seeds	86°F/68°F (30°C/ 20°C)	7 to 11	0.3 to 0.4	23 to 36
Cucumis melo	Melons	Crosser	Tender Annual	Protected Wet Seeds	86°F/68°F (30°C/ 20°C)	7 to 11	0.3 to 0.4	23 to 36
Cucumis sativus	Cucumbers	Crosser	Tender Annual	Protected Wet Seeds	86°F/68°F (30°C/ 20°C)	14 to 27	0.5 to 1	46 to 89
Cucurbita maxima	Squash (Maxima)	Crosser	Tender Annual	Protected Wet Seeds	86°F/68°F (30°C/ 20°C)	18 to 50	0.6 to 1.8	59 to 164
Cucurbita moschata	Squash (Moschata)	Crosser	Tender Annual	Protected Wet Seeds	86°F/68°F (30°C/ 20°C)	18 to 50	0.6 to 1.8	59 to 164
Cucurbita pepo	Squash (Pepo)	Crosser	Tender Annual	Protected Wet Seeds	86°F/68°F (30°C/ 20°C)	18 to 50	0.6 to 1.8	59 to 164
Solanum lycopersicum	Tomatoes	Selfer	Tender Annual	Protected Wet Seeds	86°F/68°F (30°C/ 20°C)	5 to 18	0.2 to 0.6	16 to 59
Solanum melongena	Eggplant	Intermedi-ate Selfer	Tender Annual	Protected Wet Seeds	86°F/68°F (30°C/ 20°C)	5 to 14	0.2 to 0.5	16 to 46

Index

About the Author

DAN BRISEBOIS is a founding member of Tourne-Sol Co-operative Farm, which supplies weekly farm baskets for 500 families and operates a thriving retail and wholesale seed business. He earned a B.Sc in Agricultural Engineering from McGill University and has over two decades of experience as both a market farmer and a seed producer. Dan is a founding member of the Eastern Canadian Organic Seed Growers Network, former board member of SeedChange, and past president of Canadian Organic Growers. As well as writing and teaching extensively, he mentors farmers in the art of effective planning, systems design, and time management at the Farmer Spreadsheet Academy, and hosts the popular Seed Farmer podcast. Dan is co-author of the COG Practical Skills Handbook *Crop Planning for Organic Vegetable Growers.* He lives and farms in Les Cèdres, Quebec.

ABOUT NEW SOCIETY PUBLISHERS

New Society Publishers is an activist, solutions-oriented publisher focused on publishing books to build a more just and sustainable future. Our books offer tips, tools, and insights from leading experts in a wide range of areas.

We're proud to hold to the highest environmental and social standards of any publisher in North America. When you buy New Society books, you are part of the solution!

- This book is printed on **100% post-consumer recycled paper,** processed chlorine-free, with low-VOC vegetable-based inks (since 2002).
- Our corporate structure is an innovative employee shareholder agreement, so we're one-third employee-owned (since 2015).
- We've created a Statement of Ethics (2021). The intent of this Statement is to act as a framework to guide our actions and facilitate feedback for continuous improvement of our work.
- We're carbon-neutral (since 2006).
- We're certified as a B Corporation (since 2016).
- We're Signatories to the UN's Sustainable Development Goals (SDG) Publishers Compact (2020–2030, the Decade of Action).

At New Society Publishers, we care deeply about *what* we publish—but also about *how* we do business.

To download our full catalog, sign up for our quarterly newsletter, and learn more about New Society Publishers, please visit newsociety.com.

ENVIRONMENTAL BENEFITS STATEMENT

New Society Publishers saved the following resources by printing the pages of this book on chlorine free paper made with 100% post-consumer waste.

TREES	WATER	ENERGY	SOLID WASTE	GREENHOUSE GASES
66	**5,300**	**28**	**230**	**28,700**
FULLY GROWN	GALLONS	MILLION BTUs	POUNDS	POUNDS

Environmental impact estimates were made using the Environmental Paper Network Paper Calculator 4.0. For more information visit www.papercalculator.org